Middleware Architecture

Edited by Mehdia Ajana El Khaddar

Published in London, United Kingdom

IntechOpen

Supporting open minds since 2005

Middleware Architecture
http://dx.doi.org/10.5772/intechopen.92536
Edited by Mehdia Ajana El Khaddar

Contributors
Mehdia Ajana El Khaddar, Gary S.D. Farrow, Jagdish Chandra Patni, Rochak Bajpai, Jyoti Tripathi, Sridhar Iyer, Atul Bansal, Muzaffar Rao, Thomas Newe

First published in London, United Kingdom, 2021 by IntechOpen
IntechOpen is the global imprint of INTECHOPEN LIMITED, registered in England and Wales, registration number: 11086078, 5 Princes Gate Court, London, SW7 2QJ, United Kingdom
Printed in Croatia

British Library Cataloguing-in-Publication Data
A catalogue record for this book is available from the British Library

Additional hard and PDF copies can be obtained from orders@intechopen.com

Middleware Architecture
Edited by Mehdia Ajana El Khaddar
p. cm.
Print ISBN 978-1-83969-406-6
Online ISBN 978-1-83969-407-3
eBook (PDF) ISBN 978-1-83969-408-0

For EU product safety concerns:
IN TECH d.o.o., Prolaz Marije Krucifikse Kozulić 3, 51000 Rijeka, Croatia, info@intechopen.com or visit our website at intechopen.com.

Meet the editor

Dr. Mehdia Ajana El Khaddar obtained her Ph.D. in computer science from ENSIAS School in Rabat, Morocco (Ecole Nationale Supérieure d'Informatique et d'Analyse des Systèmes) and both her Bachelor of Science in computer science and Master of Science in computer networks from Al Akhawayn University in Ifrane (AUI), Morocco. Her research fields include, but are not limited to, the following: RFID, middleware design, WSNs, pervasive computing, ubiquitous computing, context-awareness, policy-based systems, ontologies, IoT, IoE, IoNT, etc. Dr. Mehdia Ajana El Khaddar is a pioneering researcher of middleware architecture for RFID applications, WSNs, and IoT, appointed as the best reviewer by international journals and proofreading or editing companies, and holder of best paper awards from many recognized international conferences for her research papers.

Contents

Preface **XI**

Section 1
Middleware Applications **1**

Chapter 1 **3**
Middleware Solutions for the Internet of Things: A Survey
by Mehdia Ajana El Khaddar

Chapter 2 **29**
Middleware Architecture - History and Adaptation with IEEE 802.11
by Rochak Bajpai, Atul Bansal, Jyoti Tripathi and Sridhar Iyer

Chapter 3 **41**
Middleware Application, Suitable to Build an Automated and Connected Smart Manufacturing Environment
by Muzaffar Rao and Thomas Newe

Section 2
Cloud Computing Middleware **51**

Chapter 4 **53**
Middleware Patterns for Cloud Platforms
by Gary S.D. Farrow

Chapter 5 **75**
Cloud Security in Middleware Architecture
by Jagdish Chandra Patni

Preface

The idea of having this book came from my interest in producing a practical book about middleware design and architecture and gathering all the recent research studies in this field. My previous work about radio-frequency identification (RFID) middleware design and its application to many areas, such as, library management, supply chain management (SCM), and health care, has motivated me to edit this book concerning the different existing middleware design patterns and applications that have emerged recently due to the technological revolution and the increasing demand to develop smart environments.

Middleware refers to the distributed software layer that bridges the gap and removes impediments between the heterogeneous hardware platforms and the backend applications requirements. It serves as an intermediate layer providing common services and programming abstractions and hiding the low-level management of the connected hardware. With the recent advances in distributed systems and enabling technologies, such as RFID, wireless sensor networks (WSNs), internet of things (IoT), internet of energy (IoE), cloud computing, context-aware pervasive computing, ubiquitous computing, etc., middleware design and development has become a necessity, taking increasing importance. A dedicated middleware solution is required for managing and monitoring the different hardware devices, as well as processing dynamically generated high volumes of data, applying contextual rules before disseminating these data to the different connected backend applications, supporting rapid applications development, and also targeting the security, privacy, and other issues both at the hardware and applications levels.

This book provides a holistic view about the different design patterns and reference models used in middleware architectures in general, followed by a detailed survey of recent propositions of specific middleware architectures dedicated to the use of the different emerging technologies, such as, automating technologies, including but not limited to IoT, RFID, WSNs, and cloud computing.

The aim of this book is to approach middleware systems from an architectural and application perspective and to cover the middleware design and implementation challenges related to each application field. This book, therefore, intends to provide a comprehensive body of knowledge for the benefit of middleware systems' designers and developers in different application domains and to bring together in one place important and up-to-date contributions in this fast-moving research area, and also the remaining issues and challenges that still need to be considered by researchers in their future works.

The organization of this book is directed by middleware design patterns and applications' development considerations. The first section of this book presents middleware applications: Chapter 1 discusses and presents the different IoT middleware design patterns and surveys the most recent existing middleware solutions for each pattern; Chapter 2 presents the middleware architecture and its adaptation to IEEE 802.11 protocol, and Chapter 3 presents a case of middleware application used to build a connected and smart manufacturing environment. The second section of this book gives an overview of middleware solutions and design patterns for cloud platforms, which is presented in Chapter 4, followed by an overview of the concept of middleware in the context of cloud computing and a detailed discussion of the major cloud security challenges and solutions given, which is presented in Chapter 5.

Mehdia Ajana El Khaddar
Al Akhawayn University in Ifrane,
Morocco

Section 1

Middleware Applications

Chapter 1

Middleware Solutions for the Internet of Things: A Survey

Mehdia Ajana El Khaddar

Abstract

The Internet of Things (IoT), along with its wider variants including numerous technologies, things, and people: the Internet of Everything (IoE) and the Internet of Nano Things (IoNT), are considered as part of the Internet of the future and ubiquitous computing allowing the communication among billions of smart devices and objects, and have recently drawn a very significant research attention. In these approaches, there are varieties of heterogeneous devices empowered by new capabilities and interacting with each other to achieve specific applications in different domains. A middleware layer is therefore required to abstract the physical layer details of the smart IoT devices and ease the complex and challenging task of developing multiple backend applications. In this chapter, an overview of IoT technologies, architecture, and main applications is given first and then followed by a comprehensive survey on the most recently used and proposed middleware solutions designed for IoT networks. In addition, open issues in IoT middleware design and future works in the field of middleware development are highlighted.

Keywords: Internet of Things (IoT), WSNs, radio frequency identification (RFID), virtual machine, events, services, middleware architecture, context awareness, ubiquitous computing, machine-to-machine (M2M) communication

1. Introduction

Nowadays, various new generation-connected objects or things are invading our daily lives including sensors, radio frequency identification (RFID) tags, smartphones, wearables, and actuators among others, due to the emergence of new technologies. With the development of cloud computing and wireless technologies, and the emergence of new connected devices at a decreasing price, the IoT market is expected to grow rapidly fostering the development of applications in different domains, including but not limited to healthcare, manufacturing, logistics and transportation, traffic management, home automation, smart cities, smart grids, smart agriculture, etc. [1]. These applications will use the raw data generated by the different connected things/objects and provide new services in the targeted domains [2]. The Global System for Mobile Communications Association (GSMA) forecasts that "by 2025, the IoT connections will reach almost 25 billion globally" [3]. These predictions are therefore highlighting the role of IoT in providing new ways of communication over the Internet.

The IoT network is considered a heterogeneous network with a complex structure, connecting a wide range of devices using different evolving technologies such as Bluetooth, ZigBee, Wi-Fi, 3G, 4G, 5G. The ubiquitous computing environment of IoT connecting heterogeneous devices, technologies, and applications, and generating a large number of events continuously brings in important and new challenges, such as interoperability, security, confidentiality, privacy, and energy-efficient operations [4]. For example, location tracking by the IoT devices may be allowed by some people to get personalized services; however, it may violate their privacy. The middleware, which is a software application, can hide the things details from the applications by communicating with the heterogeneous connected devices/things, filtering the raw captured data, and processing them before dissemination to the connected applications, and therefore easing the backend applications' development and offering multiple common services [5]. The middleware can also deal with the interoperability, security, and privacy issues facing the IoT. The IoT middleware development is an active research area; there exist many middleware solutions addressing the IoT environment requirements in terms of context awareness, scalability, interoperability across heterogeneous things, device management, data storage and management, security, privacy, and service deployment. A major challenge faced by application developers today is finding the most appropriate IoT middleware solution in terms of the provided functionalities that should meet the application requirements and the underlying used technologies. Therefore, the existing works on IoT middleware architecture need to be analyzed to address their existing technical challenges, issues, and gaps in this domain and suggest further improvements. This chapter provides a detailed overview of existing middleware solutions for IoT and is organized as follows: Section 2 provides background about IoT characteristics, architecture, and applications, and gives an overview of the IoT middleware general architecture. Section 3 presents the IoT middleware design considerations and requirements. Section 4 provides a comprehensive review of currently existing research work in designing IoT middleware platforms. Section 5 discusses criteria for choosing the right platform according to the application requirements, along with some open issues and challenges, and the last Section 6 provides some concluding comments recommending future research directions in this area.

2. Background

2.1 IoT architecture and applications

The Internet of Things (IoT) consists of two words: The "Internet" is defined "a network of networks and a global system of interconnected computer networks that use TCP/IP as a standard Internet Protocol (IP) to connect millions of users and multiple private, public, academic, business, and government networks," and "Things" include "any real-world object/physical element such as home appliances, clothes, smartphones, etc. or living things like people, animals, or plants" [6]. The International Telecommunication Union (ITU) considers IoT as "a worldwide network of interconnected objects, allowing anything and anyone to be connected, anytime and anyplace using any network and any service" [7]. Therefore, in IoT, many heterogeneous devices will be connected to the Internet and will provide a large volume of data and even services. The major components of IoT include wireless sensors and actuators networks, machine-to-machine (M2M) communications, and RFID/near-field communication (NFC) as shown in **Figure 1**.

2.1.1 IoT infrastructure characteristics

2.1.1.1 Heterogeneous intelligent devices

In IoT, heterogeneous devices in terms of features, capacities, sensor computing natures (high end, middle end, and low end), costs, embedded intelligence

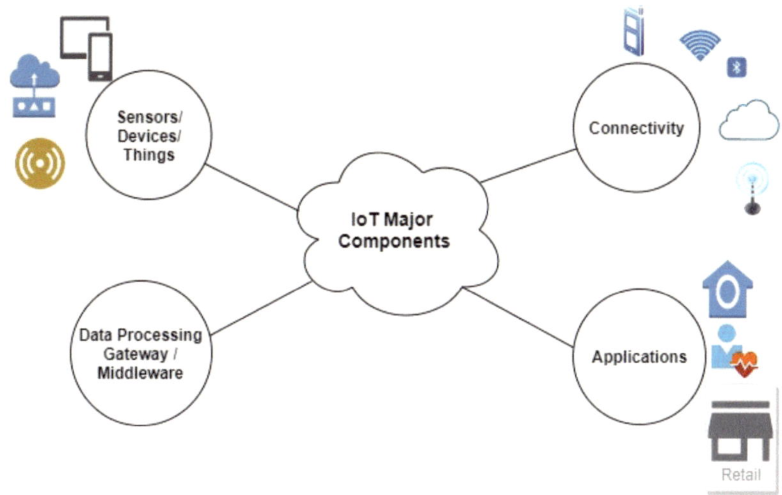

Figure 1.
IoT major components.

(adapting to the context, environment, and circumstances), and from different vendors are expected to communicate and exchange information [8]. Also, new types of devices are emerging continuously in the future as new technologies are developed [8]. **Figure 2** shows the main technologies used in IoT.

2.1.1.2 Context and location awareness

The different connected devices/things capture large volumes of data that need further processing; it should be filtered, interpreted, and put in a context to have a meaning. Context awareness helps to make the interpretation of data easier by adding context information to the raw data captured by the IoT things, which allows performing M2M communication that is considered a core element in an IoT environment [9].

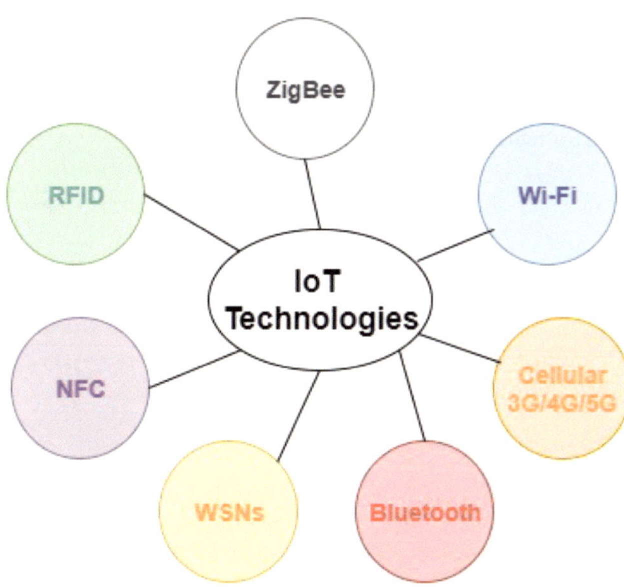

Figure 2.
IoT technologies.

Also, the spatial/location information about things is important to understand their interactions with other surrounding things (e.g., objects and people) [10].

2.1.1.3 Limited resources

IoT devices including small embedded sensors, RFID tags and readers, actuators, etc., are constrained in terms of processing, communication capacity, battery, and memory [8]. Also, the cost of these devices may increase when their performance increases in terms of processing, communication capacity, or the use of the battery to power them (e.g., active RFID tags are more expensive than the passive ones [5]).

2.1.1.4 Voluminous data and a continuous generation of spontaneous events

There are trillions of connected objects that are exchanging and storing hundreds of Exabytes of noisy data in IoT, and therefore forming an ultra-large-scale network [11]. These sudden interactions among things will also continuously generate events causing network congestion [11].

2.1.1.5 Dynamic distributed infrastructure

The IoT network is considered as an *ad hoc* network; there is no dedicated server for managing the resources of devices/things, and devices can join or leave the network anytime they want, or they can disconnect due to battery power shortage or connectivity problems. Cooperation between nodes will be needed to keep an active and stable network, and support multiple applications' development [11]. Therefore, the IoT network is considered a globally distributed network like the Internet and a local one within an application domain/context.

2.1.2 IoT applications characteristics

2.1.2.1 Diverse application domains

The IoT applications can be developed to cater to the needs of different domains and environments, having different requirements and deployment architectures, such as logistics and supply chain management, healthcare, environmental monitoring, smart home/buildings, smart agriculture [6]. **Figure 3** gives an overview of the potential IoT applications.

2.1.2.2 Real-time delivery of data and services

IoT applications in some specific domains such as transportations and healthcare need to communicate real-time data and deliver on-time services to avoid critical situations [6].

2.1.2.3 Security and privacy concerns

In the IoT network, the security of applications and communications among the different nodes should be considered, along with the privacy of people's captured data such as location, daily activities, buying habits [12]. An efficient and scalable security mechanism should be implemented considering the *ad hoc* nature of the IoT network, and also, the privacy issues should be considered not to prohibit the deployment of applications that violate citizen's privacy by the law [12].

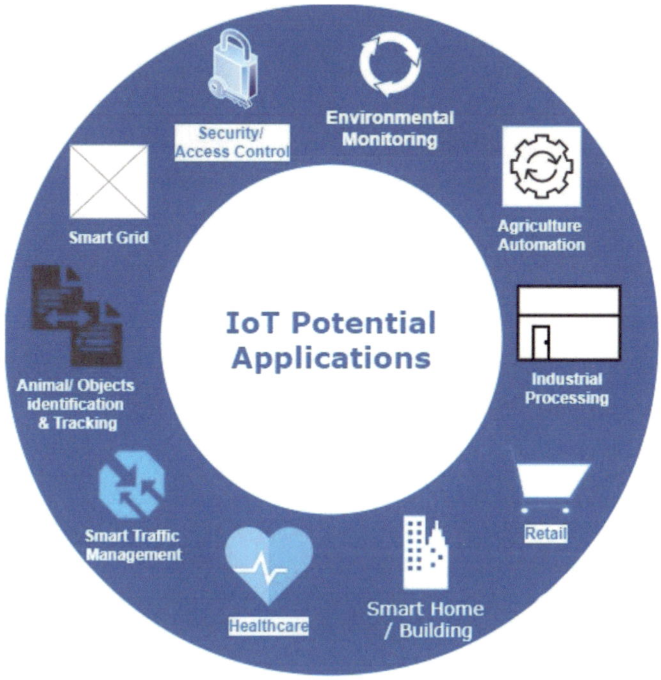

Figure 3.
IoT potential applications.

2.2 IoT middleware platform general architecture

Given the IoT infrastructure and applications' characteristics stated above, and based on my previous research work done on middleware architecture for RFID [5], context aware, and ubiquitous computing [13], an IoT middleware solution can generally provide the following functionalities:

- Device abstraction, discovery, management, and control: It includes interoperation among the heterogeneous connected devices/things using different standards. Application programming interfaces (APIs) are used for abstracting the communication with the physical layer and also for disseminating data and services to the different connected backend applications, hiding all details and complexities.

- Data management and dissemination: It provides the different data preprocessing functionalities, such as filtering, duplicate removal, aggregation.

- Context detection and processing

- Security, privacy, and business rules processing

- Application abstraction

The IoT middleware architecture is depicted in **Figure 4**. The main layers include device abstraction and resource management layer, which handle the interoperability and interaction with the heterogeneous devices, and manage the low-level hardware parameters such as the used protocols, communication technologies, standards, and air interface; data management layer is responsible for storing

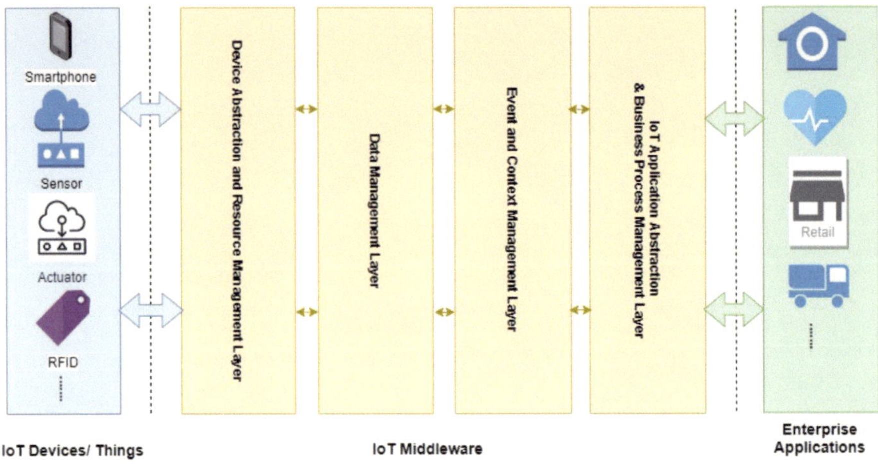

Figure 4.
IoT middleware architecture.

and processing (filtering, aggregation, inference, etc.) the raw data captured by the different devices/things; event management and context detection layer include the application of policies and business rules requested by the applications (e.g., security and privacy rules); and application abstraction layer allows the communication of applications with the different devices and helps them to get the desired processed data and generated events from the middleware.

3. IoT middleware design considerations and requirements

The role of a middleware platform is to provide a software layer shielding the complexities of the hardware layer including the operating systems from the applications and allowing the applications' developers to be concentrated mainly on the requirements/problem to be solved. As described in Section 2, in the context of IoT, there is a considerable variation in the used technologies, standards, and network communications. We describe herein, a set of design considerations and requirements for a middleware to suit the IoT infrastructure and application characteristics.

3.1 Resource discovery and management

Since the IoT infrastructure is dynamic by its nature, the IoT middleware should provide an automatic device discovery and enable the IoT heterogeneous hardware devices (e.g., RFIDs, sensors, smartphones) to detect their neighbors in the network and show their presence and available resources to them. In this case, the middleware should consider the characteristics of the resource-constrained IoT devices and be scalable in terms of the number of connected devices in the network. The middleware should also manage the devices, monitor their resource usage, and resolve any resource conflicts when potential and spontaneous new devices are connected to satisfy the application requirements.

3.2 Data management, context awareness, and event management

The IoT middleware should provide data management and processing functionalities to the backend applications; these include but are not limited to data detection and acquisition from the different connected devices/things, data preliminary processing, such as filtering, duplicate removal, compression, aggregation, and data storage. The IoT middleware should also manage the high number of generated events in an IoT environment, such as real-time dissemination of events to the applications, event transformation based on contextual/location data, and inferences.

The IoT-middleware should provide context detection and processing for it to adapt to smart applications requirements; it should collect context data and then process them to generate inferences and decisions. This could be achieved by using different techniques such as knowledge database, data mining algorithms, semantic context aware multimodal visualization approach, and the use of optimized message communication between the middleware users.

3.3 Scalability and adaptability

The IoT network can include a large number of connected things/devices and provide multiple services; therefore, the IoT middleware should be scalable allowing the growth of the IoT network, including the emergence of new heterogeneous devices that could be monitored, added, or removed without any impact on existing middleware functionalities, the provision of new services/functionalities, the addition/removal of network nodes, and the connection of multiple interesting applications in the middleware services without complexity. The use of IPv6, loose coupling, and virtualization are considered as useful ways for improving scalability in middleware solutions. Also, the use of a service-oriented architecture (SOA) makes the middleware flexible to the applications' requirements in terms of new services. The IoT middleware should also be dynamically adaptive to the different circumstances and changes in the IoT environment.

3.4 Real-time data capture and services

The IoT network deals with multiple real-time/time-critical applications requiring a timeliness delivery of processed data and services without any delay, for example, healthcare applications; therefore, the middleware should provide real-time services and information to these applications. In this case, the middleware should manage the large data volumes detected from the multiple connected devices and therefore use novel methods to detect, process, and disseminate these data to the interested applications. The challenge of transaction handling, indexing, and querying these data should also be handled. This could be ensured through the use of agents, query processors, notification managers, etc.

3.5 Reliability and availability

Every component or layer in the IoT middleware should be operational including communication, data processing, events management, technologies, devices connectivity, and application management, even when failures occur. It should provide a stable service for applications/users even at times of failure. The middleware must also be available at all times for mission-critical applications that require a high fault tolerance, for example, medical applications; in the case of failure, the recovery time should be reduced to cater to the applications' availability requirements.

3.6 Security and privacy

The IoT middleware should consider the security and privacy rules and policies required by the connected applications. The use of context awareness in the middleware can disclose some personal information about individuals such as location;

therefore, it needs to protect people's privacy using policies/rules/ontologies depending on the applications' specific needs [12]. Also, most of IoT middleware solutions are evolving into the cloud, which requires more mechanisms to be put in place to deal with the security and privacy issues, making users safe and protecting their personal data.

3.7 Ease of use and deployment

The IoT middleware should be lightweight, and easily used and deployed by the end-users of devices or applications without any complicated setup procedures.

3.8 Distributed implementation

If the IoT infrastructure is distributed, the middleware implementation should also be distributed, for example, when the devices, applications, and users are located in different geographical areas.

Some of the requirements stated above are considered to be mandatory for some applications while optional for others; for example, the real-time data capture and services are highly required in the case of medical applications, but it is optional for other applications that do not use timeliness information. However, the security, privacy, and interoperability functionalities are strictly required by all types of applications.

4. Overview of existing IoT middleware solutions

Many middleware solutions, using a single design approach (e.g., service-based, agent-based, database-based) or a hybrid one (combining different design approaches), and providing different functionalities in many application domains have been proposed and implemented in the IoT. These initiatives aim to offer a standard platform used to abstract the lower-level details of the connected physical devices and offer multiple services to the users and/or applications. In this chapter, the existing IoT middlewares are surveyed based on their used design approach and are grouped into six categories: service-oriented middleware, agent-based middleware, event-based middleware, virtual machine-based middleware, database-oriented middleware, and application-oriented middleware. A comparison of these design approaches is given in **Table 1**.

4.1 Service-oriented middleware solutions

The service-oriented middleware (SOM), based on the service-oriented design pattern, provides services to the applications, such as service discovery and management, data management, and quality of service (QoS) management. There exist many service-oriented IoT middleware solutions. Some of the commonly used service-oriented IoT middleware solutions are described as follows:

Hydra is a SOM for ubiquitous computing providing many management components for resources, security, and services [19]. Hydra is a lightweight middleware supporting dynamic self-reconfiguration and optimizing energy consumption in battery-constrained devices. The security and privacy requirements are ensured by Hydra through the use of Web Services enriched by semantic resolution [20].

The *SenseWrap* [21] middleware solution uses virtual sensors with the Zeroconf protocols to abstract the sensors' low-level details from the applications, and allow them to discover sensor-hosted services. This middleware solution applies virtualization only to sensors, which makes it unsuitable for IoT environments including heterogeneous devices and application domains.

DOI: http://dx.doi.org/10.5772/intechopen.100348

			IoT middleware requirements/features									
Middleware approach	**Middleware solutions**	**Target environment**	**Interoperability**	**Scalability**	**Adaptability**	**Real timeliness**	**Security and privacy**	**Reliability**	**Context awareness**	**Ease of use and deployment**	**Data management**	**Event management**
Service-oriented	**Hydra**	WSNs	Yes	Yes	No	Yes	Yes	No	Yes	Yes	Yes	Yes
	SenseWrap	Virtual sensors	Yes	Yes	No	No	No	No	No	Yes	No	Yes
	MUSIC	Ubiquitous	Yes	Yes	Yes	Yes	No	Yes	Yes	Yes	No	Yes
	SENSEI	Sensors/ actuators	No	Yes	Yes	Yes	Yes	No	Yes	Yes	Yes	Yes
	TinySOA	WSNs	No	Yes	Yes	Yes	No	No	No	Yes	No	No
	SensorsMW	WSNs	No	Yes	Yes	Yes	No	No	No	Yes	Yes	No
	Servilla	WSNs	No	Yes	Yes	Yes	Yes	Yes	No	Yes	No	No
	SOCRADES	Heterogeneous devices	Yes	Yes	Yes	Yes	Yes	Yes	Yes	Yes	Yes	Yes
	Middleware based on REST API	Heterogeneous devices	Yes	Yes	Yes	Yes	No	No	No	Yes	No	Yes
	3SOA	IoT	Yes	Yes	Yes	Yes	Yes	No	Yes	Yes	Yes	Yes
Cloud-based Serivce-oriented	**Google** Fit	IoT, cloud	Yes	Yes	Yes	Yes	Yes	No	Yes	Yes	Yes	No
	Xively	IoT, cloud	Yes	Yes	Yes	Yes	No	No	Yes	Yes	Yes	Yes
	CarrIoTs	IoT, cloud	Yes	Yes	Yes	Yes	Yes	Yes	Yes	Yes	Yes	Yes
	Echelon	IoT, cloud	Yes	Yes	Yes	Yes	Yes	No	No	Yes	Yes	No

			IoT middleware requirements/features									
Middleware approach	**Middleware solutions**	**Target environment**	**Interoperability**	**Scalability**	**Adaptability**	**Real timeliness**	**Security and privacy**	**Reliability**	**Context awareness**	**Ease of use and deployment**	**Data management**	**Event management**
Microservices-based	**Arrowhead Framework** [14]	IoT	Yes	Yes	Yes	Yes	Yes	No	No	Yes	Data exchange	Yes
	General microservice architecture [15]	IoT	Yes	Yes	Yes	Yes	Yes	No	Yes	Yes	Yes	Yes
	Smart City [16]	IoT	Yes	Yes	Yes	Yes	Yes	No	Yes	Yes	Yes	Yes
	Ocean [17]	IoT	Yes	Yes	Yes	yes	No	No	Yes	Yes	Yes	No
	Web of Objects Architecture [18]	IoT	Yes	Yes	Yes	Yes	No	No	Yes	yes	Yes	No
Agent-based	**Impala**	WSNs	No	No	No	Yes	No	No	No	Yes	No	Yes
	ActorNet	WSNs	No	No	No	Yes	No	Yes	Yes	Yes	Yes	Yes
	Agilla	WSNs	Yes	No	Yes	Yes	No	Yes	Yes	Yes	Yes	Yes
	Ubiware	IoT, ambient	Yes	Yes	Yes	Yes	No	No	Yes	Yes	Yes	Yes
	Smart messages	WSNs, Embedded Systems	No	No	No	Yes	Yes	No	Yes	Yes	No	No
	ACOSO-based middleware	IoT	Yes	Yes	Yes	Yes	No	Yes	Yes	Yes	Yes	Yes

			IoT middleware requirements/features									
Middleware approach	**Middleware solutions**	**Target environment**	**Interoperability**	**Scalability**	**Adaptability**	**Real timeliness**	**Security and privacy**	**Reliability**	**Context awareness**	**Ease of use and deployment**	**Data management**	**Event management**
Event-based	**EMMA**	IoT, Cloud	Yes	Yes	Yes	Yes	No	Limited	No	Yes	No	Yes
	Hermes	Large-scale distributed and ubiquitous systems	Yes	Yes	Yes	Yes	Yes	Yes	No	Yes	No	Yes
	Event-based Middleware for Syntactical Interoperability in IoT	IoT	Yes	Yes	Yes	Yes	No	Yes	No	Yes	No	Yes
Virtual-machine-Based	Maté	WSNs	No	Yes	Yes	Yes	No	No	Yes	Yes	No	No
	VM*	WSns	No	Yes	No	Yes	No	No	No	Yes	No	Yes
	Melete	WSNs	No	Yes	Yes	Yes	No	Yes	No	Yes	No	Yes
Database-oriented	**SINA**	WSNs	No	Yes	No	Yes	No	No	No	Yes	Limited	Yes
	IrisNet	WSNs	No	Yes	No	Yes	No	No	No	Yes	Yes	No
	Sensation	WSNs	No	Yes	No	Limited	No	No	No	No	Yes	No
	TinyDB	WSNs	No	No	No	Limited	No	No	No	No	Yes	No
	HyCache	WSNs	No	No	No	Limited	No	No	No	No	Yes	No

			IoT middleware requirements/features									
Middleware approach	**Middleware solutions**	**Target environment**	**Interoperability**	**Scalability**	**Adaptability**	**Real timeliness**	**Security and privacy**	**Reliability**	**Context awareness**	**Ease of use and deployment**	**Data management**	**Event management**
Application-oriented	AutoSec	Distributed	Yes	Yes	Yes	Yes	No	No	Yes	Yes	Yes	Yes
	Adaptive middleware	WSNs/ Healthcare	No	Yes	Yes	Yes	No	Yes	Yes	Yes	No	Yes
	MILAN	WSNs/ Healthcare	No	Yes	Yes	Yes	No	No	Yes	Yes	No	Yes
	MidFusion	WSNs/ Information Fusion	No	Yes	Yes	Yes	No	No	No	Yes	Yes	Yes
	TinyCubus	Driver Assistance Systems	Yes	Yes	Yes	Yes	No	No	No	Yes	Yes	No

Table 1.
Comparison of existing IoT middleware solutions.

The *MUSIC* middleware [22] supports building systems in ubiquitous environments where service providers and consumers may change dynamically based on context. Its architecture is composed of different managers providing different functionalities, including the context manager, service discovery manager, QoS manager, SLA monitoring, and adaptation manager. The use of context data by the MUSIC middleware may increase the risk of privacy leakage in an IoT environment. *SENSEI* [23] is another middleware solution including context services and a context model for the real world Internet including IoT. Its resources use ontologies for their semantic modeling, which makes it unsuitable for large-scale IoT networks since there are no standard established ontologies yet.

TinySOA [24] is a service-oriented middleware used for WSN applications development. It provides a management of WSN devices and communications, and allows applications to get processed data from the connected sensors. TinySOA allows only a few functionalities, such as abstraction and resource discovery related to WSNs, and does not support other devices; therefore, it could not be used fully within an IoT network [24]. Another SOM providing the management of quality of service in WSNs is called *SensorsMW* [25]. *Servilla* middleware also facilitates application development using heterogeneous WSNs; however, it is not widely used due to the privacy and security threats caused by the individual sensor-level access [26].

SOCRADES middleware [27] contains a layer for devices and services monitoring responsible for devices/things management and service discovery, and another one for application services such as event storage. The middleware provides a security solution by using authentication to control access to the different devices. However, the privacy of sensitive information is not ensured, since a direct access to the connected devices and their offered services is allowed by the middleware.

There exist many other cloud-based service-oriented IoT middleware solutions, such as *Google Fit*, *Xively*, *CarrIoTs*, *Echelon*; however, there are still many concerns about the cloud platform security and privacy, especially for mission-critical IoT applications [28].

Recent studies have been conducted concerning the design and implementation of service-oriented IoT middleware solutions including the one in [29] that suggests a middleware based on REST API to collect data from different devices, intending to deal with the heterogeneity issues. The authors in [30] presented a 3SOA (Sensing-as-a-Service run-time Service-Oriented Architecture) middleware solution that allows interoperability among IoT platforms, and highly abstracts the applications from the low-level details of IoT hardware platforms and communication languages.

In conclusion, most old SOMs manage only WSNs and do not scale to the use of multiple heterogeneous devices as in the context of IoT. Recent suggested service-based middleware platforms provide solutions for the interoperability and heterogeneity problems; however, they still offer a limited security through the use of authentication, do not use unified service standards, and require automation for service configuration and optimization due to the recurring demands of new services by the interesting applications.

Another type of microservices-based architecture has been recently proposed to develop IoT platforms that meet the heterogeneous and distributed nature of IoT devices, and provide dynamic, scalable, maintainable, and interoperable IoT environments. Delsing et al. [14] propose an Arrowhead Framework architecture enabling scalability, security, and real-time performance in a multi-cloud setting. This architecture supports multiple IoT devices based on SOA architecture in local clouds to exchange inter- and intra-cloud information, and allows organizations to move toward a multi-stakeholder cooperation catering to market requirements and supporting efficiency, flexibility, and sustainability [14].

A general microservice architecture for IoT applications development is proposed by Sun et al. [15], providing flexibility, scalability, maintainability, light-weightness, and loose coupling to deal with the different challenges of the continuous IoT development. The authors focus on the system design based on microservices and device communication protocols used between the service layer and physical device layer. This framework allows, therefore, more interoperability, automation, and intelligence and provides big data and geo-localization services [15].

Another recent architecture based on microservices is proposed by Lai et al. [16] to provide IoT services for multi-mobility in a smart city. The architecture provides flexibility and scalability to efficiently manage the different heterogeneous IoT devices using independent microservices, which could be separately deployed in a distributed system [16]. The authors used real-case scenarios to test the architecture using multi-mobility services for citizens in a smart city.

A recent study [17] also shows how the use of a framework based on microservices allows to mitigate the critical challenges of IoT devices and applications, and increases their scalability when deployed in the ocean where there is a continuous increasing growth of big data.

Many other microservices-based IoT platforms have been proposed in various application domains such as smart farms [31], smart logistics/factories [32], smart cars [33], and smart commerce [34]. Jarwar et al. [18] also proposed a cross-domain/general-purpose Web of Objects Architecture for IoT service provisioning in which a virtual object is used as an abstraction of a physical object.

4.2 Agent-based middleware solutions

Agent-based middleware solutions use mobile agents to facilitate distribution throughout the network and allow a partial failure tolerance. The use of mobile agents in the IoT network provides many advantages including interoperability with the heterogeneous devices, reliability and availability, resource and code management taking into consideration the resource-constrained devices, and application management. Some of the most commonly used agent-based middleware solutions are highlighted below.

Impala [35] is an agent-based middleware solution enabling code management, application modularity, resource management, mobility, and openness in WSNs. Its architecture also allows an improvement of the efficiency of resource-constrained nodes. However, Impala middleware does not provide the raw data cleaning functionality, which is necessary for an IoT setting.

Other examples of agent-based WSN middleware solutions include *ActorNet* [36] that provides context management and allows application development taking into consideration the limited resources in a WSN environment. However, ActorNet uses a service discovery mechanism leading to a slow network. *Agilla* [37] is another example of agent-based platforms, which deploys independent event-related mobile agents in every sensor node; however, this is limited due to the constrained resources of nodes, which may cause message loss and interference with programmability and code management tasks.

Ubiware [38] is considered a dedicated agent-based middleware solution for IoT, which supports resource discovery, invocation, monitoring, and the development of multiple extensible applications. Ubiware is a Java-based solution with a three-layer architecture where resources are interpreted as Java components; it uses ontologies and policies to satisfy the security and interoperability requirements; however, these policies do not include all the available WSN standards. There exist many other Java-based middleware solutions dedicated to WSN applications, such as *AFME*, *MAPS*, *MASPOT*, and *TinyMAPS* to name a few [39].

Smart messages [40] middleware is a highly flexible solution for dynamic network configurations; it overcomes the limitations of volatile, heterogeneous, and resource-constrained embedded systems using agent migration. However, it is limited in terms of the number of connected applications and its support to multiple devices in the case of an IoT context.

The authors in [41] present a new approach for increasing the smart objects' self-adaptation and allowing them to make autonomous decisions and be smarter based on a multi-agent system (MAS). The authors in [42] also presented a new multi-agent-based approach called ACOSO (Agent-based Cooperating Smart Objects) and its related middleware catering for the heterogeneous IoT platforms. The flexibility and effectiveness of this middleware were proved through the implementation of a "Smart University system."

The autonomous behavior of agents used in middleware solutions may lead to the IoT network's self-organization and fault tolerance. However, the dynamic behavior of agents may lead to message loss; therefore, most of the above-discussed middleware solutions could not be used within the large-scale IoT networks requiring a heterogeneous infrastructure, including resource-constrained devices.

4.3 Event-based middleware solutions

All the components of an event-based middleware solution use a publish/subscribe model; the event sending component is called the producer or publisher, and the receiving component is called the consumer or subscriber. The consumers are registered for a particular event published by the producers for which they are frequently receiving notifications. The event-based approach provides timeliness, security, scalability, availability, reliability, and fault tolerance.

EMMA [43] is an available Java Message Service middleware, which is a type of event-based approach designed for video communication systems to provide many types of messaging. However, it is not energy efficient and provides only a limited reliability.

Hermes middleware [44] also provides scalability, interoperability, and reliability, and it is also fault tolerant. However, it provides only a limited adaptation and does not allow a composite and persistent storage of events.

The authors in [45] proposed an event-based middleware solution implemented using the publish-subscribe pattern to solve the problem of interoperability in IoT. The interoperability assessment methodology was used to test the middleware performance, and it was shown that it is qualified compared to previous systems.

There exist many other event-based middleware solutions including *GREEN* [46], *RUNES* [47], *Steam* [48], *PSWare* [49], *PRISMA* [50], and *TinyDDS* [51], which are appropriate for systems involving a high mobility and failure occurrence. However, they do not adequately address the context awareness, adaptability, interoperability, security, privacy, and timeliness requirements of the IoT. Also, the concurrency of the event in this type of middleware solutions may lead to reduced system reliability.

4.4 Virtual machine-based middleware solutions

The virtual machine (VM) middleware approach considers virtualizing the network infrastructure, where the different network nodes are holding a VM and applications are designed as separate modules distributed throughout the network. This ensures self-management, and a high level of abstraction and adaptability. *Maté* [52] is a middleware solution based on VM, which addresses the different challenges in WSNs and is designed for nodes with limited energy and bandwidth resources. Mate is based on a VM approach and provides byte code interpretation and tackles the different challenges in WSNs; however, it does not provide event management and does not allow a single sensor node to support multiple applications. Some other middleware solutions based on the VM approach were built on top of Mate to extend its capabilities, including *VM** [53] and *Melete* [54]. These provide resource management, code dissemination, and an easy concurrent application deployment; however, they do not handle a dynamic network topology.

There exist some middleware solutions based on Java virtual machine (JVM), such as *MagnetOS*, *Squawk*, and *Sensorware* which allows them to support multiple portable applications; however, they are unsuitable for the IoT resource-constrained devices since they use heavy mechanisms for interlayer communication and computation consuming memory and processing power [55]. These constraints make the VM-based approach suitable only for resource-rich devices.

The application-specific virtual machine (ASVM) approach has been developed to target specific application domains. Middleware solutions based on this approach include but are not limited to *TinyVM* [56], *SwissQM* [57], and *TinyReef* [58]. However, the ASVM approach is still heavyweight, which makes it unsuitable for the limited-resource devices in an IoT network deployment.

4.5 Database-oriented middleware solutions

The whole network in this type of middleware solution is viewed as a relational database, managed using a query language like SQL. For example, the *Sensor Information Networking Architecture (SINA)* middleware [59] enables applications to send queries, collect results, and monitor network changes in a WSN setting. It also supports resource management and monitoring, event monitoring, data preprocessing, while clustering sensor nodes to ensure scalability and energy-efficient operations. However, SINA is not context aware, and it does not support security, privacy, and interoperability. *IrisNet* [60] is another distributed and lightweight database-oriented middleware solution providing simultaneous heterogeneous WSN services using queries over the collected data from the sensor nodes. However, it does not resolve the issues related to energy efficiency, interoperability, adaptiveness, and context awareness. Other examples of database-oriented middleware solutions include *Sensation* [61], *TinyDB* [62], and *HyCache* [63]. In these solutions, database queries are used to get approximate data of interest from the sensor nodes; they do not support the real-time requirement of the IoT infrastructure. They are also energy inefficient and use a centralized model, which does not scale to the ultra-large dynamic IoT networks [59]. Also, they do not provide the data aggregation and knowledge discovery functionalities.

4.6 Application-oriented middleware solutions

Application-oriented middleware solutions are dedicated to specific domain requirements and infrastructure. For example, the *Automatic Service Composition (AutoSec)* middleware supports one application at a time using resource provisioning and information collection policies set by the different applications [64]. *Adaptive middleware* is designed for smart home applications providing context awareness, and it also supports adaptation for other applications and ensures the quality of information collection and transmission between the network nodes [65]. Other examples include *MILAN* middleware [59] that targets the healthcare applications and adapts to their QoS requirements at runtime, *MidFusion* [66] designed for information fusion applications such as intrusion detection systems, and *TinyCubus* [59] designed for driver assistance systems that satisfies the application requirements by customizing its generic components.

The application-specific approach leads to the design of special-purpose middleware systems dedicated to a specific application domain, using a centralized mechanism for resource discovery. This makes them unsuitable for the distributed and fault-tolerant nature of IoT environments.

4.7 Hybrid approach middleware solutions

There exist some middleware platforms using a hybrid approach, combining two or more design approaches stated above. For example, both *SOCRADES* [27] and *Servilla* [26] service-oriented middleware solutions use also the virtual machine (VM)-based approach. The VM in Servilla, for example, serves to execute application tasks, while the service provisioning framework (SPF) (the service-oriented part) is used to discover and execute services on individual sensor nodes in a WSN. A middleware solution designed for the manufacturing domain using the hybrid approach is also proposed in [67], taking the advantages of both the database-oriented and semantic modeling approaches for ensuring an accurate and efficient data management and communication among the different devices and applications.

Table 1 shows the IoT requirements/features available in each middleware design approach and provides a comparison of the different IoT middleware solutions described in Section 4. The choice of the comparison criteria is based on the works cited above, from which the most common, essential, critical, and important characteristics that are shared between the different IoT platforms have been extracted. The description of each criterion is given above in Section 3 (IoT Middleware Design Considerations and Requirements). There exist many additional/non-functional criteria and features, which could be available in some IoT platforms such as recoverability, fault-tolerance, maintainability, configurability, mobility, reusability. But these are not subject of this review since it targets only the most essential design features/functionalities of IoT middleware solutions.

5. Open issues in IoT middleware design

According to the previous comparison, most of the works concentrate their efforts on providing basic functionalities such as ease of deployment, data management, event management, and real-timeliness. A considerable effort must be made in interoperability and adaptability, which allows devices/things using heterogeneous protocols to connect. Context awareness is also a feature that is not considered by most of the described middleware solutions and still encounters many shortcomings. In addition, security and privacy features need particular attention from researchers, because they are missed in almost all the reviewed middleware solutions above.

In summary, the most challenging issues that still persist in IoT-middleware design, implementation, and deployment are listed below:

- Standardization: The use of heterogeneous devices within a variety of application domains in the IoT makes the use of a single standard for a middleware solution impossible. However, many research works tend to implement a standardized middleware solution for a specific domain, such as semantic web applications domain, sensor networking environments, and smart offices [59, 65]. This will allow application developers to select a middleware solution following the desired standard within a certain domain.

- Storage capacity: The storage capacity of the heterogeneous connected things within the IoT should be considered when implementing a middleware solution. For example, if the middleware solution offers many services and data management functions, it will be difficult to use it with low-level storage devices. This issue could be addressed by defining storage requirements by the different types of backend applications, taking into consideration the application domain, before choosing an adequate middleware solution.

- Security and privacy: IoT middleware solutions can rely on a single layer for providing security and privacy to the backend applications, or distribute the security and privacy support among all the middleware layers. Either way, security and privacy support will add more processing overhead to the middleware platform, and it should also take into consideration the security and privacy requirements and rules for each specific application with minimum overhead.

- Applications abstraction: The IoT middleware should include an application abstraction layer to allow multiple backend applications to be registered with the middleware, and to specify the set of services and data processing functions needed. The applications can also specify policies/rules concerning some functionalities, such as context awareness, security, privacy, data processing, and event processing and inferences.

6. Conclusion and future work

Middleware is becoming a necessity for managing heterogeneous devices in the IoT network and developing applications in different domains. There exist a variety of middleware platforms designed for IoT. This chapter provides a detailed overview of existing IoT middleware solutions, and discusses the technical challenges and open issues involved in designing these platforms including device and application abstraction, scalability, context awareness, event management, unfixed infrastructure, security, and privacy. In future work, the open issues in IoT could be further investigated to suggest possible new approaches to solve them. Also, a new middleware design approach may be proposed to include a new perspective for managing the IoT devices/things and applications, including a solution for the unexplored open issues in a specific application domain, such as security, privacy, and interoperability. A test of this new approach could be performed using my previous proposed middleware solution for RFID described in [5].

Author details

Mehdia Ajana El Khaddar
Alakhawayn University, Ifrane, Morocco

*Address all correspondence to: mehdia.ajana@gmail.com

References

[1] Mahmoud Elkhodr M, Shahrestani S, Cheung HS. Internet of Things applications: Current and future development. In: Hassan QF, editor. Innovative Research and Applications in Next-Generation High Performance Computing. 1st ed. Hershey, Pennsylvania: IGI Global; 2016. pp. 397-427. DOI: 10.4018/978-1-5225-0287-6.ch016

[2] GLOBE NEWSWIRE. Internet of Things (IoT) Market—Growth, Trends, Forecasts (2020-2025) [Internet]. 2020. Available from: https://www.globenewswire.com/news-release/2020/05/13/2033070/0/en/The-global-IoT-market-is-expected-to-reach-a-value-of-USD-1256-1-billion-by-2025-from-USD-690-billion-in-2019-at-a-CAGR-of-10-53-during-the-period-2020-2025.html [Accessed: 01 March 2021]

[3] GSMA. IoT Connections Forecast: The Rise of Enterprise [Internet]. 2019. Available from: https://www.gsma.com/iot/resources/iot-connections-forecast-the-rise-of-enterprise/ [Accessed: 21 February 2021]

[4] Jason IH, James AL. Four Technological Challenges in Ubiquitous Computing and their Influence on Interaction Design [Internet]. Available from: https://citeseerx.ist.psu.edu/viewdoc/download?doi=10.1.1.419.5005&rep=rep1&type=pdf [Accessed: 21 February 2021]

[5] Ajana ME, Boulmalf M, Harroud H, Elkoutbi M. RFID middleware design and architecture. In: Turcu C, editor. Designing and Deploying RFID Applications. Rijeka: InTechOpen; 2011. DOI: 10.5772/16917. ISBN: 978-953-307-265-4. Available from: http://www.intechopen.com/books/designing-and-deploying-rfid-applications/rfid-middleware-design-and-architecture

[6] Ajana ME, Boulmalf M. Smartphone: The ultimate IoT and IoE device. In: Mohamudally N, editor. Smartphones from an Applied Research Perspective. Rijeka: IntechOpen; 2017. DOI: 10.5772/intechopen.69734. Available from: https://www.intechopen.com/books/smartphones-from-an-applied-research-perspective/smartphone-the-ultimate-iot-and-ioe-device

[7] Gopalsamy BN. Communication trends in Internet of Things. In: Sugumaran V, editor. Developments and Trends in Intelligent Technologies and Smart Systems. 1st ed. Hershey, Pennsylvania: IGI Global; 2018. pp. 248-305. DOI: 10.4018/978-1-5225-3686-4.ch014

[8] Yacchirema Vargas DC, Palau Salvador CE. Smart IoT gateway for heterogeneous devices interoperability. IEEE Latin America Transactions. 2016;**14**(8):3900-3906. DOI: 10.1109/TLA.2016.7786378

[9] Ntalasha D, Renfa L, Wang Y. Internet of thing context awareness model. EAI Endorsed Transactions on Context-aware Systems and Applications. 2016;**3**(7):151084. DOI: 10.4108/eai.12-2-2016.151084

[10] Cristea V, Dobre C, Pop F. Context-aware environments for the Internet of Things. In: Bessis N, Xhafa F, Varvarigou D, Hill R, Li M, editors. Internet of Things and Inter-cooperative Computational Technologies for Collective Intelligence. USA: Springer; 2013. pp. 25-49. DOI: 10.1007/978-3-642-34952-2_2

[11] Krishnamurthi R, Kumar A, Gopinathan D, Nayyar A, Qureshi B. An overview of IoT sensor data processing, fusion, and analysis techniques. Sensors. 2020;**20**(21):6076. DOI: 10.3390/s20216076

[12] Oorschot P C Van, Smith S W. The Internet of Things: Security challenges.

IEEE Security & Privacy. 2019;**17**(5):7-9. DOI: 10.1109/MSEC.2019.2925918

[13] Ajana ME, Chraibi M, Harroud H, Boulmalf M, Elkoutbi M, Maach A. FlexRFID: A security and service control policy-based middleware for context-aware pervasive computing. International Journal of Advanced Research in Artificial Intelligence (IJARAI). 2014;**3**(10):26-34. DOI: 10.14569/IJARAI.2014.031004

[14] Delsing J et al. The arrowhead framework architecture: Arrowhead framework. In: Delsing J, editor. IoT Automation. United States: CRC Press Publisher; 2017. DOI: 10.1201/97813 15367897-4. ISBN: 9781498756754

[15] Sun L, Li Y, Memon RA. An open IoT framework based on microservices architecture. China Communications. 2017;**14**(2):154-162. DOI: 10.1109/ CC.2017.7868163

[16] Lai C, Boi F, Buschettu A, Caboni R. IoT and microservice architecture for multimobility in a smart city. In: Proceedings of the IEEE 7th International Conference on Future Internet of Things and Cloud (FiCloud); 26-28 August 2019; Istanbul, Turkey. New York: IEEE; 2019. pp. 238-242. DOI: 10.1109/FiCloud.2019.00040

[17] Razzaq A. Microservices architecture for IoT applications in the Ocean: Microservices architecture based framework for reducing the complexity and increasing the scalability of IoT applications in the Ocean. In: Proceedings of the 20th International Conference on Computational Science and Its Applications (ICCSA); 1-4 July 2020; Cagliari, Italy. New York: IEEE; 2020. pp. 87-90. DOI: 10.1109/ICCSA50381. 2020.00025

[18] Jarwar MA, Kibria MG, Ali S, Chong I. Microservices in web objects enabled IoT environment for enhancing reusability. Sensors. 2018;**18**(2):352. DOI: 10.3390/s18020352

[19] Eisenhauer M, Rosengren P, Antolin P. HYDRA: A development platform for integrating wireless devices and sensors into ambient intelligence systems. In: Giusto D, Iera A, Morabito G, Atzori L, editors. The Internet of Things. New York: Springer; 2010. pp. 367-373. DOI: 10.1007/978-1-4419-1674-7_36

[20] Reiners R, Zimmermann A, Jentsch M, Zhang Y. Automizing home environments and supervising patients at home with the hydra middleware: Application scenarios using the hydra middleware for embedded systems. In: Proceedings of the First International Workshop on Context-aware Software Technology and Applications; 24 August 2009; Amsterdam, The Netherlands. New York: ACM; 2009. pp. 9-12. DOI: 10.1145/1595768.1595772

[21] Zgheib R, Conchon E, Bastide R. Semantic middleware architectures for IoT healthcare applications. In: Ganchev I, Garcia N, Dobre C, Mavromoustakis C, Goleva R, editors. Enhanced Living Environments. Cham: Springer; 2019. pp. 263-294. DOI: 10.1007/978-3-030-10752-9_11

[22] Rouvoy R, et al. MUSIC: Middleware support for self-adaptation in ubiquitous and service-oriented environments. In: Cheng BHC, de Lemos R, Giese H, Inverardi P, Magee J, editors. Software Engineering for Self-Adaptive Systems. Berlin: Springer; 2009. pp. 164-182. DOI: 10.1007/978-3-642-02161-9_9

[23] Tsiatsis V et al. The SENSEI real world internet architecture. In: Georgios T, et al., editors. Towards the Future Internet—Emerging Trends from European Research. Amsterdam, The Netherlands: IOS Press; 2010. pp. 247-256. DOI: 10.3233/978-1-60750-539-6-247

[24] Avilés-López E, García-Macías JA. TinySOA: A service-oriented

architecture for wireless sensor networks. Service Oriented Computing and Applications. 2009;**3**:99-108. DOI: 10.1007/s11761-009-0043-x

[25] Anastasi G F, Bini E, Lipari G. Extracting data from WSNs: A service-oriented approach. In: Anastasi G, Bellini E, Di Nitto E, Ghezzi C, Tanca L, Zimeo E, editors. Methodologies and Technologies for Networked Enterprises. Berlin: Springer; 2012. p. 329-356. DOI: 10.1007/978-3-642-31739-2_17

[26] Chien-Liang F, Gruia-Catalin R, Chenyang L. Servilla: A flexible service provisioning middleware for heterogeneous sensor networks. Science of Computer Programming. 2012;**77**(6):663-684. DOI: 10.1016/j.scico.2010.11.006

[27] de Souza LMS, Spiess P, Guinard D, Köhler M, Karnouskos S, Savio D. SOCRADES: A web service based shop floor integration infrastructure. In: Floerkemeier C, Langheinrich M, Fleisch E, Mattern F, Sarma SE, editors. The Internet of Things. Berlin: Springer; 2008. pp. 50-67. DOI: 10.1007/978-3-540-78731-0_4

[28] Ngu AH, Gutierrez M, Metsis V, Nepal S, Sheng QZ. IoT middleware: A survey on issues and enabling technologies. IEEE Internet of Things Journal. 2017;**4**(1):1-20. DOI: 10.1109/JIOT.2016.2615180

[29] Mesmoudi Y et al. A Middleware based on service oriented architecture for heterogeneity issues within the Internet of Things (MSOAH-IoT). Journal of King Saud University—Computer and Information Sciences. 2020;**32**(10):1108-1116. DOI: 10.1016/j.jksuci.2018.11.011

[30] Hammoudeh M et al. A service oriented approach for sensing in the Internet of Things: Intelligent transportation systems and privacy use cases. IEEE Sensors Journal. 2020; **21**(14):15753-15761. DOI: 10.1109/JSEN.2020.2981558

[31] Taneja M et al. SmartHerd management: A microservices-based fog computing-assisted IoT platform towards data-driven smart dairy farming. Software Practice and Experience. 2019;**49**:1055-1078. DOI: 10.1002/spe.2704

[32] Herrera-Quintero LF et al. Smart ITS sensor for the transportation planning using the IoT and Bigdata approaches to produce ITS cloud services. In: Proceedings of the IEEE 8th Euro American Conference on Telematics and Information Systems (EATIS); 28-29 April 2016; Cartagena, Colombia. New York: IEEE; 2016. pp. 1-7. DOI: 10.1109/EATIS.2016.7520096

[33] Kanti Datta S et al. IoT and microservices based testbed for connected car services. In: Proceedings of the IEEE 19th International Symposium on "A World of Wireless, Mobile and Multimedia Networks" (WoWMoM); 12-15 June 2018; Chania, Greece, Piscataway, NJ: IEEE; 2018. pp. 14-19. DOI: 10.1109/WoWMoM.2018.8449768

[34] Banerjee A, Jiang B. A Blockchain-based IoT platform integrated with cloud services. In: Proceedings of the International Conference on Parallel and Distributed Processing Techniques & Applications; July 29th - August 1st 2019; Las Vegas, Nevada. C. S. R. E. A., 2020.

[35] Liu T, Martonosi M. Impala: A middleware system for managing autonomic, parallel sensor systems. ACM SIGPLAN Notices. 2003;**38**(10):107-118. DOI: 10.1145/966049.781516

[36] Kwon Y, Sundresh S, Mechitov K, Agha G. ActorNet: An actor platform for wireless sensor networks. In:

Proceedings of the 5th International Joint Conference on Autonomous Agents and Multiagent Systems; 8-12 May 2006; Hakodate, Japan. New York: ACM, 2006. DOI: 10.1145/1160633.1160871

[37] Fok CL, Gruia-Catalin Roman GC, Lu C. Agilla: A mobile agent middleware for self-adaptive wireless sensor networks. ACM Transactions on Autonomous and Adaptive Systems. 2009;**4**(3):1-26. DOI: 10.1145/1552297.1552299

[38] Vasile-Marian Scuturici VM, Surdu S, Yann G, Petit JM. UbiWare: Web-based dynamic data & service management platform for AmI. In: Proceedings of the Posters and Demo Track Conference; 3 December, 2012; Montreal Quebec Canada. New York: ACM; 2012. DOI: 10.1145/2405153.2405164

[39] Aiello F, Fortino G, Galzarano S, Vittorioso A. TinyMAPS: A lightweight java-based mobile agent system for wireless sensor networks. In: Proceedings of the 5th International Symposium on Intelligent Distributed Computing (IDC 2011); October 2011; Delft, the Netherlands: Springer-Verlag Berlin Heidelberg; 2012. DOI: 10.1007/978-3-642-24013-3_16

[40] Kang P et al. Smart messages: A distributed computing platform for networks of embedded systems. The Computer Journal. 2004;**47**(4). DOI: 10.1093/comjnl/47.4.475

[41] Chekati A, Riahi M, Moussa F. Agent-based modelling approach for decision making in an IoT framework. In: Barolli L, Woungang I, Enokido T, editors. Advanced Information Networking and Applications. AINA; 2021. Lecture Notes in Networks and Systems, vol 226. Springer, Cham. DOI: 10.1007/978-3-030-75075-6_21

[42] Fortino G et al. An Agent-Based Middleware for Cooperating Smart Objects. In: Proceedings of the 11th International Conference on Practical Applications of Agents and Multi-Agent Systems; 22-24; May, 2013; Salamanca, Spain: Springer-Verlag Berlin Heidelberg; 2013. pp. 387-398. DOI: 10.1007/978-3-642-38061-7_36

[43] Rausch T, Nastic S, Dustdar S. EMMA: Distributed QoS-aware MQTT middleware for edge computing applications. In: Proceedings of the IEEE International Conference on Cloud Engineering (IC2E); 17-20 April, 2018; Orlando, FL, USA: IEEE; 2018. pp. 191-197. DOI: 10.1109/IC2E.2018.00043

[44] Pietzuch PR. Hermes: A scalable event-based middleware. University of Cambrige Computer Laboratory Technical Report N° 590; 2004. ISSN 1476-2986. Available from: https://www.cl.cam.ac.uk/techreports/UCAM-CL-TR-590.pdf

[45] Pramukantoro ES, Anwari H. An event-based middleware for syntactical interoperability in Internet of Things. International Journal of Electrical and Computer Engineering. 2018;**8**(5):3784. DOI: 10.11591/ijece.v8i5.pp3784-3792

[46] Sivaharan T, Blair G, Coulson G. Green: A configurable and reconfigurable publish-subscribe middleware for pervasive computing. In: Meersman R, Tari Z, editors. On the Move to Meaningful Internet Systems. Berlin: Springer; 2005. pp. 732-749. DOI: 10.1007/978-3-540-78731-0_4

[47] Costa P, et al. The runes middleware for networked embedded systems and its application in a disaster management scenario. In: Proceedings of the IEEE 5th Annual International Conference on Pervasive Computing and Communication (PerCom'07); 19-23 March 2007; White Plains, NY, USA: Computer Society; 2007; pp. 69-78. DOI: 10.1109/PERCOM.2007.36

[48] Meier R, Cahill V. Steam: Event-based middleware for wireless ad hoc networks. In: Proceedings of the IEEE

22nd International Conference on Distributed Computing Systems Workshops; 2-5 July 2002; Vienna, Austria: IEEE; 2002. pp. 639-644. DOI: 10.1109/ICDCSW.2002.1030841

[49] Lai S, Cao J, Zheng Y. Psware: A publish/subscribe middleware supporting composite event in wireless sensor network. In: Proceedings of the IEEE International Conference on Pervasive Computing and Communication (PerCom'09); 9-13 March 2009; Galveston, TX, USA: IEEE Computer Society; 2009. pp. 1-6. DOI: 10.1109/PERCOM.2009.4912862

[50] Silva JR, et al. PRISMA: A publish-subscribe and resource-oriented middleware for wireless sensor networks. In: Proceedings of the 10th Advanced IEEE International Conference on Telecommunications; 20-24 July 2014; Paris, France. International Academy, Research, and Industry Association (IARIA); 2014. pp. 87-97

[51] Boonma P, Suzuki J. TinyDDS: An interoperable and configurable publish/subscribe middleware for wireless sensor networks. In: Hinze A, Buchmann A, editors. Principles and Applications of Distributed Event-Based Systems. Hershey, Pennsylvania: IGI Global; 2010. p. 206. DOI: 10.4018/978-1-60566-697-6.ch009

[52] Levis P, Culler DE. Maté: A tiny virtual machine for sensor networks. In: Proceedings of the Tenth ACM International Conference on Architectural Support for Programming Languages and Operating Systems; 5-9 October 2002; San Jose, California, United States: ACM; 2002. DOI: 10.1145/605406.605407

[53] Koshy J, Pandey R. Vm: Synthesizing scalable runtime environments for sensor networks. In: Proceedings of the 3rd International Conference on Embedded Networked Sensor Systems (SenSys '05); 2-4 November 2005; San Diego, California, USA: ACM; 2005. pp. 243-254. DOI: 10.1145/1098918.1098945

[54] Khalid Z, Fisal N, Rozaini M. A survey of middleware for sensor and network virtualization. Sensors. 2014;**14**(12):24046-24097. DOI: 10.3390/s141224046

[55] Costa N, Pereira A, Serodio C. Virtual machines applied to WSN's: The state-of-the-art and classification. In: Proceedings of the Second International Conference on Systems and Networks Communications (ICSNC 2007); 25-31 August 2007; Cap Eterel, France, IEEE Computer Society; 2007. pp. 50-50. DOI: 10.1109/ICSNC.2007.83

[56] Hong K et al. Tinyvm: An energy-efficient execution infrastructure for sensor networks. Software: Practice and Experience. 2012;**42**(10):1193-1209. DOI: 10.1002/spe.1123

[57] Mueller R, Alonso G, Kossmann D. SwissQM: Next generation data processing in sensor networks. In: Proceedings of the Third Biennial Conference on Innovative Data Systems Research (CIDR); 7-10 January 2007; Asilomar, CA, USA. Online Proceedings. Available from: www.cidrdb.org 2007. pp. 1-9. DOI: 10.3929/ethz-b-000004843

[58] Marques IL, Ronan J, Rosa NS. TinyReef: A register-based virtual machine for Wireless Sensor Networks. In: Proceedings of the IEEE International Conference on SENSORS; 25-28 October 2009; Christchurch, New Zealand: IEEE; 2009. pp. 1423-1426. DOI: 10.1109/ICSENS.2009.5398437

[59] de Freitas EP. A Survey on Adaptable Middleware for Wireless Sensor Networks. Halmstad University Technical Report IDE0851; 2008. Available from: http://www.diva-portal.org/smash/get/diva2:239429/FULLTEXT01.pdf

[60] Deshpande A, Suman N, Gibbons PB, Seshan S. IrisNet:

Internetscale Resource-Intensive Sensor Services. In: Proceedings of the ACM SIGMOD International Conference on Management of Data; 9-12 June 2003; San Diego, California, USA: ACM; 2003. p. 667. DOI: 10.1145/872757.872856

[61] Hasiotis T et al. Sensation: A middleware integration platform for pervasive applications in wireless sensor networks. In: Proceedings of the IEEE Second European Workshop on Wireless Sensor Networks; 31 January - 2 February 2005; Istanbul, Turkey: IEEE; 2005. pp. 366-377. DOI: 10.1109/EWSN.2005.1462028

[62] Madden S, Franklin MJ, Hellerstein JM, Hong W. TinyDB: An acquisitional query processing system for sensor networks. ACM Transactions on Database Systems. 2005;**30**(1):122-173. DOI: 10.1145/1061318.1061322

[63] Zhao D, Raicu I. HyCache: A user-level caching middleware for distributed file systems. In: Proceedings of the IEEE International Symposium on Parallel & Distributed Processing, Workshops and PhD Forum(IPDPSW); 20-24 May 2013; Cambridge, MA, USA: IEEE; 2013. pp. 1997-2006. DOI: 10.1109/IPDPSW.2013.83

[64] Han Q, Venkatasubramanian N. Autosec: An integrated middleware framework for dynamic service brokering. IEEE Distributed Systems Online. 2001;**2**(7):22-31

[65] Huebscher MC, McCann JA. Adaptive middleware for context aware applications in smart-homes. In: Proceedings of the 2nd Workshop on Middleware for Pervasive and Ad-hoc Computing; 18-22 October 2004; Toronto, Ontario, Canada, United States: ACM; 2004. pp. 111-116. DOI: 10.1145/1028509.1028511

[66] Alex H, Kumar M, Shirazi B. MidFusion: An adaptive middleware for information fusion in sensor network applications. Information Fusion. 2008;**9**(3):332-343. DOI: 10.1016/j.inffus.2005.05.007

[67] Grevenitis K et al. A hybrid framework for industrial data storage and exploitation. Procedia CIRP. 2019;**81**:892-897. DOI: 10.1016/j.procir.2019.03.221

Chapter 2

Middleware Architecture - History and Adaptation with IEEE 802.11

Rochak Bajpai, Atul Bansal, Jyoti Tripathi and Sridhar Iyer

Abstract

Communication, which intends to provide a link between any two people, is now moving towards man-to-machine and machine-to-machine connection for transferring different types of data. This transmission scenario, with and ever expanding number of active and passive users, lays the foundation to variety of communication protocols owing to the different types of data which is involved in the process. Within this ever expanding communication arena, Middle-ware can be thought of as a set of hardware and software which is used to connect different platforms with the end-users that are increasing in number day-by-day, with a possible wide spread over any region spanning from few meters to several kilometers. IEEE 802.11 is the set of standards which guides the wireless technology for device implementation and demands seamless integration across the entire protocol stack. This in turn demands an overview of the middleware architecture in broader perspective. This chapter explores the concept of middleware in the existing communication scenario, current trends and future scope.

Keywords: Middleware architecture, Communication protocols, IEEE 802.11, Network layer, Application layer, TPM, RPC, MOM, ORB

1. Introduction

Widespread usage of communication technology, intended to transfer information from one person to another, is moving rapidly towards information exchange between man and machine. The advent of computer along with digital formatted communication in light of advanced algorithms, low cost VLSI (Very Large Scale Integration) and high efficiency in computation paves the way for massive information exchange among individuals.

Simultaneously, ubiquitous computing paradigm with sensor based data communication presents unique man-to-machine information exchange which not only spread across large area but is also found to be a source of voluminous data, that needs to be carefully segregated and segmented for figuring out intelligence from a large pile of existing data.

Conventionally, middleware is said to be collection of algorithms, components and devices which enables the information exchange between different entities. As per the OSI model, middleware can be placed between the transport and the application layer, as shown in **Figure 1**.

In existing literature, the concept of middleware was discussed starting in the year 1968 in a report of the North Atlantic Treaty Organization [1], where it was placed between the application programs and the service routines. This paradigm

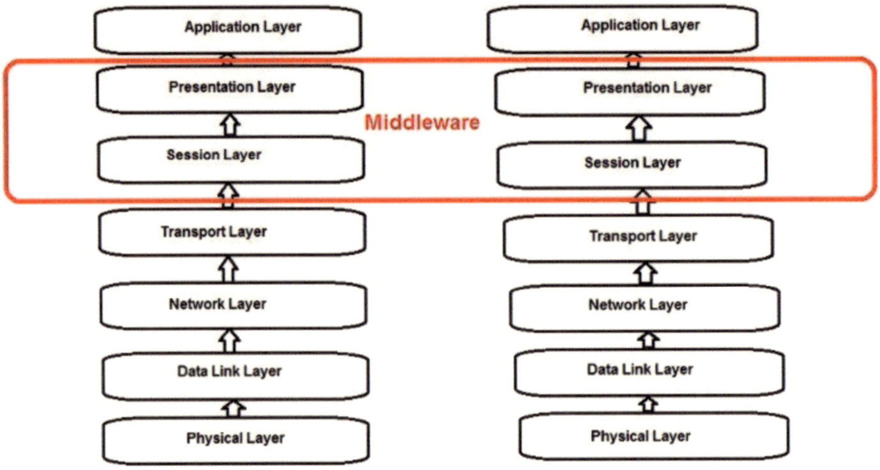

Figure 1.
Location of middleware in OSI reference model.

was quite pragmatic due to its ability to interconnect new components with the existing hardware within the same distributed system.

With the advent of the recent 5G technologies along-with strong data networking capability, a plethora of innovative sensor-based application are appearing over the horizon. This motivates us to look at the middleware architecture at a microscopic level as it involves different sensors with their applications without considering the interconnection involved i.e. it hides the heterogeneous nature of the underlying sensing data to yield the support system with the help of unified interface for the final application. This results in an easy access of the network by sensors as well as different applications, which can call the sensing network with the middleware open interface. This helps in the reduction of the design and the development cost with improved efficiency.

2. History and motivation

The initial development of the computer was based on the premise and promise of high-speed calculations which was iterative in nature. It was initially proposed through mechanical gear system but was found to be excessively time consuming. With the advent of vacuum tubes, the hurdle of timing is attended to and further, investigation in the field of solid state electronics which results in devices such as, diodes, transistors and operation amplifiers paves way for low power consuming, portable size devices which are termed as analog computers. This was the first era of computers [2]. The more robust, fast and complex Integrated Circuits (IC) were responsible for the design and development of digital computers which are having tremendous capability to perform high speed and complex calculations [3].

With the elaboration of theoretical concepts and their implementation in terms of ideas such as, distributed computing, abstraction, object-oriented programming, etc., results in the coding of complex software programs to an easily solvable problem at hand. At the same time, in order to reduce the design and development cost of the software and its interaction with the hardware, the concept of re-usability comes into software design paradigm [4].

The parallel progress in networking paradigm, which was simply started with possible interconnection of machines, later resulted in the development of OSI model to yield a layered structure of communication among different machines [5]. The layered architecture is profoundly based on the software abstraction model which felicitates a very specific section/property of the network to change without hampering the process of communication among other machines through different layers. This paradigm was the primary basis of middle ware architecture which was placing it between the transport and the application layer, as shown in **Figure 1**.

With the expansion of networking paradigm, Middleware architecture found its place among the widely used concepts, especially in the era of plug and play design methodology. In the subsequent section, we will look at the different attributes which Middleware workable.

2.1 Features of middleware

As discussed in the introduction, middleware is a set protocol for data exchange between the transport and the application layer. It helps different machines to share the data with the network and vice versa without considering the heterogeneity of the underlying data from both sides, and results in reduced development cost and improvement in the efficiency [6].

The following features are the major attributes of the middleware architecture:

1. From the Software management perspective, Middleware is found to be more and more integrated into the operating system, which results in the application evolved in a machine to be safely ported on the network as well, wherein the network is acting as an up-scaled version of the machine. Also, this results in resource management mechanism based on service quality and the flexible configuration capability [7].

2. With the widespread availability of the Internet, middleware architecture provides web-based services and resource sharing capacity, making middleware architecture almost like connecting glue which supports in the running of the application software successfully. Also, the resource sharing methods based on Internet services are more universal, cost effective and efficient.

3. Middleware architecture has transformed the conventional spoke and wheel system into the distributed system by combining different technologies such as, cloud computing, big data and virtualization, which provides the capacity to integrate different resources and yields more robust service capability. With this architecture, one can solve the issue of data storage, processing and transmission among the different internetworked systems.

Depending on the different attributes, the middleware architecture can be classified into four types [8, 9].

1. Transaction Processing Monitors (TPM)

2. Remote Procedure Call (RPC)

3. Message-Oriented Middleware (MOM)

4. Object Request Brokers (ORB)

These types are based on the varied services offered by the specific process.

Transaction Processing Monitors (TPM) is designed to monitor the successful transactions from one stage to another stage. In case of any error, TPM takes an appropriate action to rectify the error. A TPM supports optimal resource sharing among the applications with the following functionalities:

- Monitoring operation/transactions
- Managing queues
- Coordination among resources
- Creating new processes on requirement
- Secure access to services
- Wrapping data messages into messages
- Unwrapping messages into data packets
- Handling errors
- Hiding the details of inter-process communications from programmer

Remote Procedure Call (RPC) is an inter-process communication facility generally used in a client server based model with the following functionalities:

- Supports process-oriented or thread-oriented models
- Hiding details of inter-process communications from programmer
- Useful in local environment as well as distributed environment
- Performance improvement can be achieved by omitting unwanted protocol layers

Message-Oriented Middleware (MOM) is an asynchronous technique that passes the messages between transmitting and receiving application with a communication channel. Its asynchronous nature makes the applications decoupled from each other as MOM is responsible for message management system.

Object Request Brokers (ORB) acts like a broker between a client request for a service from a distributed object and the completion of that request with following functionalities:

- Life cycle service
- Persistence service
- Naming service
- Event Service
- Concurrency control service

- Transaction service
- Relationship service
- Externalization service
- Query service

In view of the above-mentioned services offered by the different types of middleware, following are the major attribute of services offered by any middleware structure [8]:

- **Presentation management**: Forms manager, graphics manager, hypermedia linker, and printing manager.
- **Computation**: Sorting, math services, internationalization services (for character and string manipulation), data converters, and time services.
- **Information management**: Directory server, log manager, file manager, record manager, relational database system, object-oriented database system, repository manager.
- **Communications**: Peer-to-peer messaging, remote procedure call, message queuing, electronic mail, electronic data interchange.
- **Control**: Thread manager, transaction manager, resource broker, fine grained request scheduler, coarse-grained job scheduler.
- **System management**: Event notification service, accounting service, configuration manager, software installation manager, fault detector, coordinator, authentication service, auditing service, encryption service, access controller.

3. IEEE 802.11 adaptation with middleware

The emerging trend of wireless technology and the associated innovative applications has changed the communication landscape drastically. Now, the reduced cost of data, high computation power of smartphones and the 5G enabled sensor technology is the major driving force behind the widespread adaptation of network services such as, map enabled movement, shipment tracking, and interactive gaming such as, Pokemon.

Inherently, multimedia services are found to be wideband in nature which, with the unpredictable channel characteristic of wireless medium, place a challenging condition to maintain reliable communication with a predefined Quality of Service (QoS).

IEEE 802.11 standard, proposed by IEEE for local area network (LAN) protocol, specifies the physical layer (PHY) and media access control (MAC) protocols for the implementation of wireless local area network (WLAN) communication in the frequency bands such as, 2.4 GHz, 5 GHz, 6 GHz, and 60 GHz. Presently, various tributaries of IEEE 802.11 are framing our day-to-day communication across the world. As per the convention, IEEE 802.11 standards are defined for physical and data link layer of the OSI data communication network protocol and become the de facto standard for wireless communication.

Moreover, the heterogeneous nature of different services such as, Wi-max, Wi-Fi, Bluetooth, and many more, with their unique data link layer presentation have their own quality control mechanism for data transmission. This results in a complex connection strategy management at network, data link, and physical layer.

In comparison, at the application layer, the QoS management parameters such as, semantic, presentation etc., have their own constraints to be followed. These constraints may not be followed every time due to resource limitations at the physical layer or multiplexing issue due to the fluctuating number of users sharing the resource pool in the end system and the network.

In the process of communication, wireless channel demands adaptive QoS implementation for peer-to-peer communication, due to its dynamic behavior. This places a bound on the middleware to manage the communication as it connects the lower layers of the communication protocol (which actually perform the communication at the data level) to the application layer (which is responsible for communication in the correct semantic and form) [10, 11]. Thus, the unique position of middleware demands a monitoring as well as an adaptive perspective such that the desired QoS requirements can be maintained at the application layer without disturbing the underneath communication.

This QoS maintenance requires a two-fold strategy, first is the monitoring of application performance and the second is adaptation of service to maintain pre-specified quality of service [11]. This adaptation strategy requires a perfect synchronization of middleware level with lower level of protocols.

These conditions lay the foundation for the middleware control framework. The middleware control framework has three primary concerns to address [12]:

- Coordinate the adaptation of all the concurrent application tasks in the end system globally i.e., maintain fairness in the architecture.

- Increase the adaptation effectiveness to maintain the QoS.

- To monitor on-the-fly dynamics in the heterogeneous environment to achieve optimum control over the smooth functioning of the applications.

3.1 IEEE 802.11 - an interesting journey

IEEE 802 project was aimed to establish the standards for physical layer (PHY) and medium access control (MAC) layer to support deployment of the local area network. The first candidate was IEEE 802.3, popularly known as Ethernet. It was a wired connection mechanism to connect devices based on carrier sense multiple access with collision detection (CSMA/CD) mechanism. Its wide acceptability with industry and home users, motivates the researcher to look for its replication in wireless domain as well which yields the IEEE 802.11 standard [13, 14].

Due to inherent unpredictable nature of radio/wireless medium paves the way for numerous deliverables of IEEE 802.11 standards, which have been flourished over the last twenty years [15, 16], such as 802.11a, 802.11e, 802.11f, 802.11 t and so on. Readers are encouraged to refer [17] to explore the need and their solutions under IEEE 802.11 horizon.

3.2 Middleware in IEEE 802.11 environment

Under the middleware architecture a brief account of various cases has been summarized below.

Authors in [18] address the issue of mobility management in increasing integration of Internet with telecommunication network which give rise to distributed computing environment. Authors proposed three mobile computing services by incorporating user virtual environment (UVE), mobile virtual terminal (MVT), and virtual resource management (VRM) based on mobility middleware solution for mobile agent.

In [19], the Authors illustrate Quality of Service (QoS) maintenance for multimedia applications over wireless network which are characterized by their limited bandwidth. Authors proposed a novel two level QoS architecture by providing service differentiation at network level and service adaptation at the middleware level. Authors validate their results with experiments and show that specific QoS levels for multimedia applications can be optimally achieved in IEEE 802.11 based wireless network.

Costa et al. in their path breaking work proposed the real time-WiFi architecture to address the issues faced by IEEE 802.11 networks in high density industrial environment [20]. Authors compare the performance of proposed architecture against the standard distributed coordination function (DCF), point coordination function (PCF), hybrid coordination function (HCF) controlled channel access (HCCA) and enhanced distributed channel access (EDCA) medium access control mechanism. Authors used a realistic error-prone model to monitor the impact of message losses in the real time Wi-Fi architecture and their result confirm that the proposed architecture performs better as compared to existing IEEE 802.11 standard mechanisms. Their architecture also offers almost consistent access delay which is one of the major requirements for real time applications.

Authors in [21] use context meta-information to improve the system performance by describing a context management middleware that can successfully handle context irrespective of the execution environment's heterogeneity.

Cruz et al. in [22], reviewed the middleware framework for Internet of Things (IoT) from software perspective. Authors rigorously explored the existing literature and analyzed the reference model for IoT platforms and proposed the basic security feature for this software. Authors also detailed the difficulties in achieving and enforcing a universal standard for middleware in the IoT structure.

Authors in [23], present a fundamental analysis to quantify the goodput performance parameter for IoT. Authors show the closed form expression of goodput as a function of the data payload length, frame retry count, data rate of transmission and wireless channel condition. Authors proposed a novel link adaptation scheme for MAC protocol data units for known wireless channel model.

Increasing use of mobile devices for location sharing applications such as Google map, OLA, Uber, pose a challenge of maintaining adequate user privacy with location sharing services and exchange of information across high heterogeneity among connecting technologies and devices. Authors in [24] proposed a middleware prototype to answer these challenges with two level proxy-based architecture as a solution.

Hamidreza et al. in [25], proposed service-oriented architecture for middleware to resolve the issue of heterogeneity among various sensors in IEEE 802.15 based wireless sensor network.

Authors in [26] discuss different types of sensor network applications with overview of related middleware and infer that none of the existing approaches can provide all the management tools required by sensor network applications. Authors showcase their new middleware MILAN with sensor-based health monitoring system.

In [27], the authors explored the concept of inter-vehicular communications. The field of vehicle to infrastructure and vehicle to vehicle communications were undertaken as well. This work provides detailed account of underlying technology under each layer with rich resource references.

Authors in [28], address the combination of vehicular ad-hoc networks (VANETs) with the social Internet of things (SIoT). Their work describes two fold relations which can be established between the vehicles and between the vehicle and road side unit (RSU). Authors proposed a social Internet of vehicle middleware to incorporate the functionalities of the intelligent transportation systems station architecture, defined by ISO and ETSI standards to integrate VANET with SIoT. They present their proof of concept with simulation results.

Pease et al. in [29] present an adaptive middleware methodology to provide robust mission critical/ military communication by providing timely MANET communications with predictive selection and dynamic contention reduction, without going for invasive protocol modification. To address the issue they proposed a novel Real-time Optimized Ad hoc Middleware based architecture (ROAM). They demonstrate the adaptability, scalability of the architecture along-with the capability to bound maximum delay, jitter and packet loss in complex and dynamic MANET's with extensive simulations.

In [30], the authors elaborated a novel mobile collaboration architecture (MoCA), a service oriented middleware architecture which support the development and deployment of distributed context-aware applications for mobile users. Authors explained the compatibility of proposed MoCA with existing software engineering principles responsible for design and implementation of context aware applications. Authors also present different prototype applications that have been developed on the top of MoCA.

Authors in [10] addressed end to end delays and security issues in application implementation. Authors proposed an integrated solution with middleware adaptation to provide tunable delay and security support according to network condition. To Support the proof of concept, authors perform test-bed experiments to showcase successful meeting of delay and security requirements in IEEE 802.11 based wireless environment.

4. Conclusion

Middleware architecture defines the connection protocol between the network layer and the application layer. In view of the on-going advances in the mobile communication, there is a requirement of a better understanding about Middleware functionality with IEEE 802.11 protocols which are responsible for the design rules of modern wireless communication.

In this chapter, we addressed the key functionalities of the middleware architecture in addition to its adaptation to the IEEE 802.11 protocols.

In the current scenario, where the need of ubiquitous connectivity is reality, the need of minimum end to end delay with almost no loss in data i.e. maintaining stringent Quality of Service at every end, is a challenging task.

This monograph will be helpful for the researchers to investigate middleware architecture in IEEE 802.11 framework to deliver robust design solution in wireless network.

Conflict of interest

"The authors declare no conflict of interest."

Author details

Rochak Bajpai[1*], Atul Bansal[2], Jyoti Tripathi[3] and Sridhar Iyer[4]

1 KIET Group of Institutions, Delhi, NCR, India

2 GLA University, Mathura, India

3 G.B. Pant Government Engineering College, Okhla, New Delhi, India

4 S.G. Balekundri Institute of Technology, Belagavi, India

*Address all correspondence to: rochakbajpai@gmail.com

References

[1] Bauer F.L., Bolliet L., Helms H. J., "Software Engineering: Report on a conference sponsored by the NATO science committee", Garmisch, Germany, Tech. Rep. Jan. 1969. \url{ http://homepages.cs.ncl.ac.uk/brian.randell/NATO/nato1968.PDF} [Accessed: 04 January 2021]

[2] Mindell DA. "Between human and machine: feedback, control, and computing before cybernetics". JHU Press; 2002 Oct 11.

[3] Graham S, Baliga G, Kumar PR. "Abstractions, architecture, mechanisms, and a middleware for networked control". IEEE Transactions on Automatic Control. 2009 Jun 30;54(7):1490-503.

[4] Prieto-Diaz R. "Status report: Software reusability". IEEE software. 1993 May;10(3):61-6.

[5] Kumar S, Dalal S, Dixit V. "The OSI model: Overview on the seven layers of computer networks". International Journal of Computer Science and Information Technology Research. 2014 Jul;2(3):461-6.

[6] Dong H. "Middleware Technologies". Journal of Anhui Vocational College of Electronics \& Information Technology. 2006:04.

[7] Lu ZJ, Yan Z. "Middleware Technology Research and Interface Design in Ubiquitous Network". In2015 Seventh IEEE International Conference on Measuring Technology and Mechatronics Automation 2015 Jun 13 (pp. 659-662).

[8] Bernstein PA. "Middleware: a model for distributed system services". Communications of the ACM. 1996 Feb 1;39(2):86-98.

[9] Tian Y, Geiger JV, Su H, Kumar SV, Houser PR. "Middleware-based sensor web integration". IEEE Journal of Selected Topics in Applied Earth Observations and Remote Sensing. 2010 Jun 21;3(4):467-72.

[10] Li B, Nahrstedt K. "A control-based middleware framework for quality-of-service adaptations". IEEE journal on selected areas in communications. 1999 Sep;17(9):1632-50.

[11] Ong CS, Xue Y, Nahrstedt K. "A middleware for service adaptation in differentiated 802.11 wireless networks". InProceedings. 2004 12th IEEE International Conference on Networks (ICON 2004) (IEEE Cat. No. 04EX955) 2004 Nov 19 (Vol. 1, pp. 364-368). IEEE.

[12] Li B, Nahrstedt K. A control-based middleware framework for quality-of-service adaptations. IEEE journal on selected areas in communications. 1999 Sep;17(9):1632-50.

[13] Tanenbaum AS, Wetherall D. Computer networks. Prentice-Hall international editions. 1996 Mar:I-XVII.

[14] Peterson LL, Davie BS. Computer networks: a systems approach. Elsevier; 2007 Apr 16.

[15] Stallings W. Local networks. ACM Computing Surveys (CSUR). 1984 Mar 29;16(1):3-41.

[16] Stallings W. Data and computer communications. Pearson Education India; 2007.

[17] Hiertz GR, Denteneer D, Stibor L, Zang Y, Costa XP, Walke B. The IEEE 802.11 universe. IEEE Communications Magazine. 2010 Jan 22;48(1):62-70.

[18] Bellavista P, Corradi A, Stefanelli C. Mobile agent middleware for mobile computing. Computer. 2001 Mar;34(3):73-81.

[19] Ong CS, Xue Y, Nahrstedt K. A middleware for service adaptation in differentiated 802.11 wireless networks. InProceedings. 2004 12th IEEE International Conference on Networks (ICON 2004)(IEEE Cat. No. 04EX955) 2004 Nov 19 (Vol. 1, pp. 364-368). IEEE.

[20] Costa R, Lau J, Portugal P, Vasques F, Moraes R. Handling real-time communication in infrastructured IEEE 802.11 wireless networks: The RT-WiFi approach. Journal of Communications and Networks. 2019 May 29;21(3):319-34.

[21] da Rocha RC, Endler M. Middleware: Context management in heterogeneous, evolving ubiquitous environments. IEEE Distributed Systems Online. 2006 May 15;7(4).

[22] da Cruz MA, Rodrigues JJ, Al-Muhtadi J, Korotaev VV, de Albuquerque VH. A reference model for internet of things middleware. IEEE Internet of Things Journal. 2018 Jan 23;5(2):871-83.

[23] Qiao D, Choi S, Shin KG. Goodput analysis and link adaptation for IEEE 802.11 a wirelessLANs. IEEE transactions on Mobile Computing. 2002 Oct;1(4):278-92.

[24] Bellavista P, Corradi A, Giannelli C. Efficiently managing location information with privacy requirements in wi-fi networks: a middleware approach. In2005 2nd International Symposium on Wireless Communication Systems 2005 Sep 5 (pp. 91-95). IEEE.

[25] Abangar H, Barnaghi P, Moessner K, Nnaemego A, Balaskandan K, Tafazolli R. A service oriented middleware architecture for wireless sensor networks. InProceedings of future network and mobile summit conference 2010 Jun.

[26] Heinzelman WB, Murphy AL, Carvalho HS, Perillo MA. Middleware to support sensor network applications. IEEE network. 2004 Jun 28;18(1):6-14.

[27] Jawhar I, Mohamed N, Zhang L. Inter-vehicular communication systems, protocols and middleware. In2010 IEEE Fifth International Conference on Networking, Architecture, and Storage 2010 Jul 15 (pp. 282-287). IEEE.

[28] Nitti M, Girau R, Floris A, Atzori L. On adding the social dimension to the internet of vehicles: Friendship and middleware. In2014 IEEE international black sea conference on communications and networking (BlackSeaCom) 2014 May 27 (pp. 134-138). IEEE.

[29] Pease SG, Phillips IW, Guan L. Adaptive intelligent middleware architecture for mobile real-time communications. IEEE Transactions on Mobile Computing. 2015 Apr 27;15(3):572-85.

[30] Viterbo J, Sacramento V, Rocha R, Baptista G, Malcher M, Endler M. A middleware architecture for context-aware and location-based mobile applications. In2008 32nd Annual IEEE Software Engineering Workshop 2008 Oct 15 (pp. 52-61). IEEE.

Chapter 3

Middleware Application, Suitable to Build an Automated and Connected Smart Manufacturing Environment

Muzaffar Rao and Thomas Newe

Abstract

The current manufacturing transformation is represented by using different terms like; Industry 4.0, smart manufacturing, Industrial Internet of Things (IIoTs), and the Model-Based enterprise. This transformation involves integrated and collaborative manufacturing systems. These manufacturing systems should meet the demands changing in real-time in the smart factory environment. Here, this manufacturing transformation is represented by the term 'Smart Manufacturing'. Smart manufacturing can optimize the manufacturing process using different technologies like IoT, Analytics, Manufacturing Intelligence, Cloud, Supplier Platforms, and Manufacturing Execution System (MES). In the cell-based manufacturing environment of the smart industry, the best way to transfer the goods between cells is through automation (mobile robots). That is why automation is the core of the smart industry i.e. industry 4.0. In a smart industrial environment, mobile-robots can safely operate with repeatability; also can take decisions based on detailed production sequences defined by Manufacturing Execution System (MES). This work focuses on the development of a middleware application using LabVIEW for mobile-robots, in a cell-based manufacturing environment. This application works as middleware to connect mobile robots with the MES system.

Keywords: MES, Robots, Cell-based manufacturing, Middleware, ROS

1. Introduction

Initially, the drive of automation was to upturn productivity and to decrease the cost associated with human operators. But now the automation also emphasizes to improve flexibility and quality in a manufacturing process [1]. Automation is an essential part of the Industry 4.0 strategy, which is fully reshaping the old manufacturing process and supporting the growth of the business. Industrial automation refers to a mechanism that combines hardware and software. Industrial automation is the use of control systems and information technologies to handle different types of machinery and processes in an industry to replace a human being. Robotic automation [2–4] is used in several areas of manufacturing industries. Robots can be used to perform tasks like assembly, welding, shipping, product packing, and handling raw materials. Many manufacturers are leveraging robotic automation for many applications. Robotic automation offers manufacturers growing opportunities and to remain competitive. Currently available industrial robots are multi-functional, so a single robot can be used for several different tasks. Sharing of the same workspace for robots and humans is becoming necessary in a smart manufacturing environment. To achieve this, robots should have the capabilities to sense and communicate. Modern mobile-robots (AIVs) can move autonomously from one place to another to achieve defined goals. For two decades, mobile-robots are in continuous research, and the rapid advances in robotics, computer vision & artificial intelligence are resulting in robots that can potentially hear and see more precisely than humans. This advancement in mobile-robots demands their rapid deployment in an industrial environment. The manufacturing process is evolving and from the robotics aspect, the industries are moving from Autonomous Guided Vehicles (AGVs) to Autonomous Intelligent Vehicles (AIVs). The use of AIVs is also reshaping the production lines in a smart factory.

A production line is a configuration of a factory that consists of a sequence of manufacturing steps. The current ideology of manufacturing depends on the linear production line. This ideology works in case of high volume demand of identical goods. But, the linear production line is not the most efficient way in the case of a large volume of goods with several choices. Manufacturers are looking for a cell-based approach to offer products with different varieties. As the cell-based approach involves more complex production flows so for this, conveyor line production systems are not suitable [5]. In cell-based manufacturing, the best way to handle the goods shifting between cells is the use of mobile-robots. Intelligent mobile-robot that knows the environment in which it operates, and that can calculate the optimized route between different cells are the best to use in the above mentioned cell-based manufacturing system. The linear and cell-based production line concept in a factory that manufactures PCBs is shown in **Figure 1**.

This work is about the development of a middleware application using LabVIEW software to control/manage the mobile-robot. This middleware

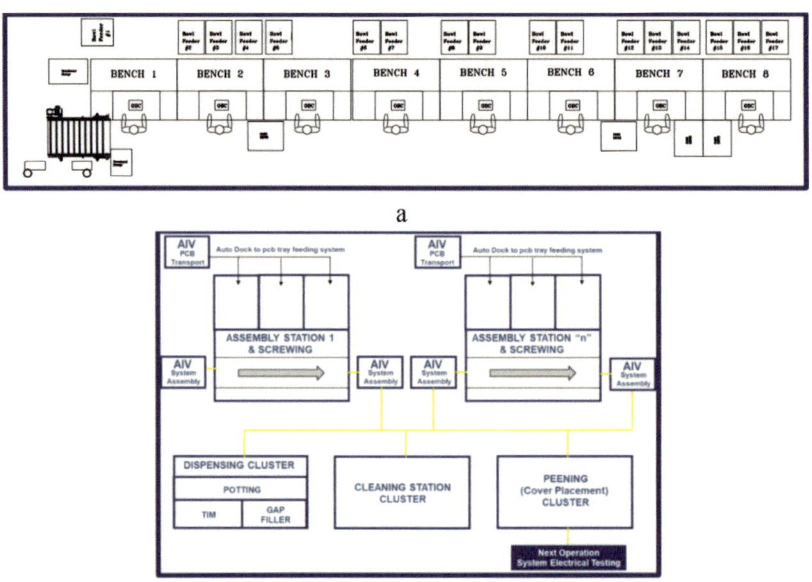

b

Figure 1.
Linear vs. cell-based production line concept. (a) Traditional linear production line. (b) Cell-based production line.

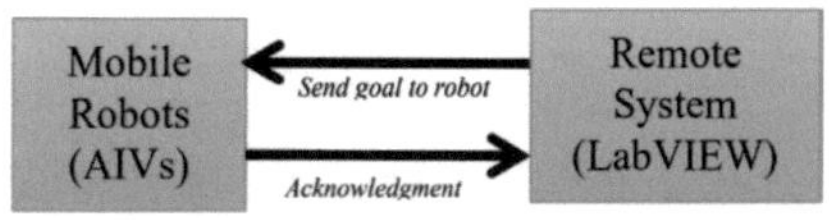

Figure 2.
Controlling of mobile-robots from a remote system using LabVIEW.

application is installed on the remote system mentioned in **Figure 2**. This work is part of an ongoing project, which involves the integration of fleets of mobile-robots and Manufacturing Execution System (MES) [6–8]. Here, the focus is on the development/testing of the LabVIEW application that can communicate with mobile-robots. The LabVIEW software is used by keeping in mind the future advancement of this application, which needs communication of fleets of robots with MES and this can be achieve using LabVIEW as middleware. The designed software application can be used with any ROS (Robot Operating System) based Robot. Here, turtlebot-3 [9, 10] and RB-1 (Robotnik) [11] are used for real-time testing.

This paper is organized as follows: Section 2 covers the industrial automation details and Section 3 discusses Mobile Robots (AIVs). Section 4 is about Robot Operating System (ROS) and Section 5 provides details of the target application. In Section 6 development of the LabVIEW application is given while Section 7 provides a discussion and Section 8 concludes.

2. Industrial automation

Manufacturers continuously face pressure to increase productivity and as mentioned in the introduction section, automation is the key to enable manufacturers to move closer to that goal in addition to meet the flexibility and quality requirements. There are numerous advantages of industrial automation, out of which some are given as [12]:

2.1 Quality control

Quality control is needed to build the customers' trust. A high level of quality compliance can be achieved by using robots in a manufacturing environment.

2.2 Repeatability

To be sure about consistency and the same quality end product is difficult without automation and this is achievable using repeatability. While, repeatability can be achieved using robots that are capable to perform the exact similar task in exactly the same way, repeatedly. This can help to save time as fewer errors result in less wasted time.

2.3 Waste reduction

The repeatability (mentioned above) allows manufacturers to minimize overall waste. Less error, as a result of repeatability, helps not only to save time, but it also minimizes the amount of material needed to yield the product. For example, robots can use less wire for welding, cut closer to the edge, and can use less amount of paint.

2.4 Faster cycle times

The production cycle time can be greatly improved using a robot. The more production helps to get the higher demand and ultimately brings more money.

2.5 Improved workplace

Safety Robots can be used in places which are not suitable for human. So, automation also helps to build safety environment.

2.6 Reduction of labor costs

Robots can replace jobs by removing workers from tough roles, which can help to reduce labor costs.

2.7 Reduced floor space

It's easy to start spreading out across the shop floor with extra materials, machinery, and tools. Robots can help to reduce the footprint of the required workspace by optimizing everything into a smaller, confined space.

2.8 Integration with business systems

Machinery and Robots can talk with each other to give a better view of the operating environment. This can help to make a smart decision on how to improve their process.

The next section describes mobile-robots with the focus on the robotic platform used here for testing.

3. Mobile-robots (AIVS)

Almost half a century ago, the first simple robot stepped onto the factory floor [13]. Today, a modern manufacturing business can be hardly imagined without the involvement of robots. The integration of robots with smart technologies allows the creation of more independent robotic systems that are not only able to carry out basic repetitive operations such as assembling, loading or modifying parts, but can also perform cognitive tasks, improving processes without human intervention and making instant decisions. In this work, turtlebot-3 and RB-1 (Robotnic) robots are used for the real-time testing of the developed LabVIEW application.

Turtlebot is a robot that is based on ROS Standard. There are three versions of turtlebot, turtlebot-1, turtlebot-2 and turtlebot-3 released in 2010, 2012, and 2017 respectively. There are two models of turtlebot-3 namely burger and waffle. Here, the burger model of turtlebot-3 is used, which is an affordable, small, and the ROS based mobile-robot. The core technology of the turtlebot-3 is SLAM (Simultaneous Localization and Mapping), Navigation, and Manipulation. Turtlebot-3 uses a Laser Distance Sensor (LDS-01), which is a 2D laser scanner capable of sensing 360 degrees that collects a set of data around the robot to use for SLAM. The turtlebot is connected/controlled through a remote system using wifi connection. The designed LabVIEW software is also tested with RB-1, which is used for real-time implementation in a factory environment. RB-1 is an autonomous and configurable robot, focused on the field of research in indoor applications. RB-1 is also based on ROS and uses an open architecture and modular control. The RB-1 platform comes with

Figure 3.
Turtlebot-3(burger) and RB-1(Robotnik).

a Hokuyo URG-04LX-UG01 [14], a 2D laser range finder for navigation. Turtlebot-3(burger) and RB-1 are shown in **Figure 3**.

The mobile-robots used in this work are based on Robot Operating System (ROS), which detail is given in the next section.

4. Robot operating system (ROS)

ROS is a meta-operating system for robots [15]. This is an open-source platform which functions are equivalent to what we expect from an operating system. These functions include low-level device control, hardware abstraction, message passing between processes, implementation of commonly-used functionality, and package management. The ROS meta operating system (shown in **Figure 4**) also provides libraries and tools for obtaining, building, writing, and running code across multiple computers. This meta-operating system is different than conventional operating systems in a way that it can be used for a different combination of hardware implementation.

However, unlike conventional operating systems, it can be used for numerous combinations of hardware implementation. Furthermore, it is a robot software platform that gives many development environments specialized for developing robot application programs. ROS runs on Ubuntu operating system; here we used Ubuntu 16.40 LTS. ROS supports many programming languages; here we used C++. There are different versions of ROS; here ROS Kinetic is used. In ROS, the smallest running unit of the processor is called a node. The first step is to run ROS_Master. Upon startup of the master, a node registers information such as name, message type, URI address, and port number of the node. The registered node can act as a publisher or subscriber based on the registered information, and nodes can exchange messages using topics and services. In this work, as mentioned earlier, the software application is developed using LabVIEW, which has support for ROS. The ROS for LabVIEW is a set of LabVIEW VIs that enables two-way communication

Figure 4.
Meta-operating system [15].

between ROS (running on Ubuntu machine) and LabVIEW (running on a windows machine). The LabVIEW ROS allows users to initialize nodes, publish and subscribe to various types of topics, and creates a ROS_Master within LabVIEW.

The next section mentioned target application details, which will help to understand the overall project concept.

5. Target application

This work is part of an ongoing project related to the integration of a fleet of mobile-robots and MES, targeted to design for the cell-based manufacturing environment. The target plant used traditionally linear manufacturing production lines for the assembly of its products. This approach takes a continuous flow in-line with balanced operations. The concern with this approach moving forward is the high up-front investment cost to achieve a Return On Investment (ROI). The line is designed for one product type, it is not flexible and is unsuitable for new product demand due to new product business or product revisions that may arise. The development of new smart manufacturing assembly lines i.e. cell-based will allow for flexibility to volume fluctuations, will support product ramp up/down, and the use of the latest automation technologies such as Industry 4.0 and the Internet of Things (IoT). The smart manufacturing assembly line will be used for multiple products, having the capability of a flexible production system for all production operations.

In cell-based manufacturing, different cells will be connected by an automatic product delivery system which will transport the product through the manufacturing process from process cell to process cell as defined by detailed production sequences using an MES. The automatic product delivery system will take the form of a fleet of AIVs with the ability to collect and deliver products to process cells in a format suitable for product feed. As mentioned earlier, this work focuses on one part of this target application i.e. development of a LabVIEW application that can communicate between the fleet of robots and MES. Details of this development are given in the next section.

6. Development of LabVIEW application

To prepare the ROS package for mobile-robots, ROS Kinetic is installed on Ubuntu 16.04 system and a code is prepared in C++ to publish and subscribe messages. This code declares a subscriber that subscribes to a topic (published by LabVIEW publisher) to receive commands for the robot and publishes a topic (subscribed by LabVIEW subscriber) to send an acknowledgment. Therefore, we can say that here publisher/subscriber sends/receives data to/from LabVIEW respectively. To take coordinates of physical location a map is generated using SLAM. Initially, the robot is taken around the workplace and allowed to scan the surrounding area with its main LiDAR sensor. It stitches that information together to form a complete static map of the workplace. The robot uses the map to calculate the best route between any two points. Here we are using hard codded coordinates of physical locations. The coordinates are taken using amcl node, which takes laser scans, transforms messages and outputs estimated pose.

The LabVIEW code consists of two while loops; one for publisher and the other for a subscriber. The ROS node is initialized using ROS_Topic_init subVI. The input of this subVI is connected with topic name, topic type, action (publisher or subscriber), and update rate and queue size. The node name (/LV1) is also assigned using the 'node' terminal of ROS_Topic_init.vi. The ROS_Topic_init.vi is connected to ROS_Topic_Read.vi (in case of subscriber) and ROS_Topic_Write.vi (in case of publisher). The 'msg_in' input of ROS_TOPIC_Write.vi is connected with 'msg_out' of add_string.vi. The message which needs to publish is connected with the 'String' terminal of add_String.vi. While, in subscriber the 'Reply' terminal of ROS_TOPIC_Read.vi is connected with 'msg-in' of parse_string.vi. The string output of the parse_string.vi is connected as an indicator, which shows the robot response. The subscriber and publisher topics are closed using ROS_Topic_Close.vi. Multiple instances of the LabVIEW application are used to control more than one Mobile-Robots.

The developed software application is initially tested using turtlebot-3, as mentioned in Section 3. For this testing, two Linux systems and one windows system are used. Out of two Linux systems, one is used as remote-PC (Ubuntu 16.04) and the other is referring to turtlebot-PC (Raspbian). The ROS packages and all dependent packages are installed on remote PC and turtlebot-PC. All steps mentioned in [16] are followed to setup turtlebot, remote-PC, and turtlebot-PC. The Windows PC is used to run the LabVIEW. ROS communicate between systems using IP addresses; so, the turtlebot-PC, remote-PC, and windows-PC connected to the same wifi router. The .bashrc file is edited on turtlebot-PC and remote-PC in such a way that ROS_Master runs on remote-PC. The map is generated using SLAM node and then we noted coordinates of target physical locations. These coordinates are needed for ROS packages. Testing is started by running ROS_Master on remote-PC and then bring-up packages of turtlebot are executed. Navigation node runs as a next step (on remote-PC), which opens the rviz (visualization tool) with the selected map. The map shows turtlebot at some random point, so the estimated posture of robot is set in rviz. Upon running the ROS node, it keeps looking for a message from LabVIEW. The LabVIEW application pops-up a window to enter the Master_IP_address that is the address of remote-PC where ROS_Master is running. LabVIEW application sends some predefined strings that receive by subscriber running on remote-PC. Based on the received strings, target location coordinates are transferred to turtlebot. Upon reaching the target location, turtlebot sends an acknowledgment. This shows a successful communication between LabVIEW and mobile-robots. The same testing is done using RB-1, only the difference is that here we need only two systems (LabVIEW system and RB-1 system) as in this case ROS_Master runs on RB-1.

This work provides the foundation to develop a software application to control mobile- robots that can be used to integrate a fleet of mobile-robots to the MES system. Direct communication between MES and the mobile-robots is not possible that's why we introduced the middleware LabVIEW application and presented this idea in [17], which summary is given below.

Previously presented work [17] emphases on the integration of an MES and AIV that is needed to develop a completely automated, connected smart manufacturing environment. This integration requires a middleware application to execute the command translations between MES and AIVs. Here, a LabVIEW based application is built as the middleware. The middleware application development is divided into 03 major parts: (1) LabVIEW & MES communication, (2) LabVIEW & AIV communication, and (3) scheduler. In part (1), the middleware application imports relevant webservices of MES and produce corresponding LabVIEW VIs. The produced LabVIEW VIs are configured using the LabVIEW. NET functions. In part (2), the middleware application utilizes the 'ROS for LabVIEW' (LabVIEW add-on) to establish communication with an AIV. Part (3) comprises of a full cycle of scheduler operation, that consists of eight steps. These eight steps are: checking of pickup locations status, selection of one pickup location, pick the product, confirm to the MES about completion of pickup operation, checking of drop-off locations status, selection of one drop-off location, drop the product, and confirm to the MES about the completion of the drop-off operation. The middleware application was tested in two stages, during the development phase, the Turtlebot-3 robot was used for testing and finally, the middleware application was commissioned using the Robotnik 'RB-1'robot in an actual factory environment. The developed middleware application supports 03 pickup and 03 drop-off locations. The pickup and drop-off locations are chosen based on pre-defined rules. The interface of the developed middleware application can be easily used by a non-technical operator and it displays a live log of operation.

The presented concept can be expended at a fleet level using the following: (a) Communication between multiple instances of LabVIEW application with each robot on the factory floor – this is a decentralized approach and the problem with this approach is the risk of assigning the same tasks at the same time to more than one robot. But this approach is ideal when each robot is assigned a dedicated zone in a manufacturing environment (b) Running ROS_Master using LabVIEW application and control all robots in the field – The issue with this centralized approach is that there will be no communication with robots in the field in case if the system running LabVIEW stops working. This chapter focused on the case (a). The LabVIEW application will take the instructions from MES and send the command to robots in the field accordingly. The LabVIEW-MES communication is not the focus of this chapter.

7. Conclusion

This work is part of an on-going project which involves the integration of a fleet of mobile-robots (AIVs) with MES. This integration can be achieved using LabVIEW software as middleware because it can support both LabVIEW-to-Robots and LabVIEW-to-MES communication. This work focuses on the development/ testing of LabVIEW-to-Robots communication. In the final setup, the LabVIEW system will work as a middleware between multiple Robots and MES. The designed LabVIEW application will be further developed to meet the overall requirements of fleet management. The updated version of the application will allow more complex production flows and the process of calling robot to pick the load; will be more intelligent by linking with MES in a fleet environment. This will provide an even larger boost to productivity, flexibility, and quality in a smart manufacturing environment.

Acknowledgements

This publication has emanated from research supported in part by a grant from Science Foundation Ireland under Grant numbers 13/RC/2094 and 16/RC/3918.

Conflict of interest

The authors declare no conflict of interest.

Author details

Muzaffar Rao and Thomas Newe
Centre for Robotics and Intelligent Systems (CRIS), Department of Eelectronic and Computer Engineering, University of Limerick, Ireland

*Address all correspondence to: muzaffar.rao@ul.ie

References

[1] M. Wiktorsson , A. Granlund, M., Lundin, B. Södergren, "Automation and Flexibility: Exploring Contradictions in Manufacturing Operations" Conference: 23rd EurOMA conference, Trondheim, Norway

[2] R. Szabó and A. Gontean, "Industrial robotic automation with Raspberry PI using image processing," 2016 International Conference on Applied Electronics (AE), Pilsen, 2016, pp. 265-268

[3] S. Choi, W. J. Eakins and T. A. Fuhlbrigge, "Trends and opportunities for robotic automation of trim & final assembly in the automotive industry," 2010 IEEE International Conference on Automation Science and Engineering, Toronto, ON, 2010, pp. 124-129

[4] B. Chu et al., "Robotic automation system for steel beam assembly in building construction," 2009 4th International Conference on Autonomous Robots and Agents, Wellington, 2009, pp. 38-43

[5] B. Shishir, (2010). Changing from conveyor to work-cell-based systems to deal with fluctuations in demand. Proceedings of the Institution of Mechanical Engineers, Part B: Journal of Engineering Manufacture, 224(1), 169-176. URL: https://doi.org/10.1243/09544054JEM1374

[6] Hennessey, T. MES is an Essential Part of Smart Manufacturing URL: https://www.ibaset.com/blog/mes-is-an-essential-part-of-smartmanufacturing/(online)

[7] MESA International. MES Explained: A High Level Vision for Executives.Available online: https://services.mesa.org/resourcelibrary/showresource/334444c5-388f-4360-beb4-3c86dc0f4de4

[8] SAP MES. Available online: https://www.sap.com/uk/produc ts/execution-mes.html.

[9] D. Singh, E. Trivedi, Y. Sharma and V. Niranjan, "TurtleBot: Design and Hardware Component Selection," 2018 International Conference on Computing, Power and Communication Technologies (GUCON), Greater Noida, Uttar Pradesh, India, 2018, pp. 805-809.

[10] Turtlebot (online), URL: https://www.turtlebot.com/

[11] Robotnik(RB-1) (online), URL: https://www.robotnik.eu/manipulators/rb-one/

[12] RobotWorx, "Advantages of Industrial Automation with Robots" Robotic Articles(online), URL: https://www.robots.com/articles/advantages-of-industrial-automationwith-robots

[13] SaM Solutions, "Five Reasons to Implement Robotics in Manufacturing" (online), URL: https://www.samsolutions.com/blog/category/manufacturing/

[14] Hokuyo "Distance Data Output/URG-04LX-UG01"(online), URL: https://www.hokuyo-aut.jp/search/single.php?serial=166

[15] Y. Pyo, H. Cho, R. Woon J., T. Lim, "ROS Robot Programming " First Edition Dec 22, 2017 Published by ROBOTIS Co.,Ltd ISBN 979-11- 962307-1-5

[16] Robotics e-manual Turtlebot-3 (online), URL: http://emanual.robotis.com/docs/en/platform/turtlebot3/overview/

[17] M. Rao; L. Lynch; J. Coady; D. Toal; T. Newe, "Integration of an MES and AIV Using a LabVIEW Middleware Scheduler Suitable for Use in Industry 4.0" Applications. *Appl. Sci.* 2020, *10*, 7054. https://doi.org/10.3390/app10207054

Section 2

Cloud Computing Middleware

Chapter 4

Middleware Patterns for Cloud Platforms

Gary S.D. Farrow

Abstract

This chapter explores how traditional system architectures are being affected by the emergence of 'Uber' style platform models that provide business services with huge global reach. The specific demands and characteristics of such platforms are discussed which in turn dictate their technical requirements. The chapter will explain how middleware technologies have evolved to support today's requirements for such massively scalable platform solutions. The latest preferred architectural paradigms dictate the use of micro-services and APIs are central to the design of such platforms. Similarly, event based architectures are another key paradigm that must be supported. The role of modern middleware and cloud technologies to support these newly dominant paradigms will be explained. Key architectural patterns pertinent to global platform solutions are illustrated. The role of modern middleware in fulfilling these patterns is highlighted using real-world examples from the field of open finance.

Keywords: Cloud architecture, migration patterns, API ecosystem, event-based architecture, microservices, cloud migration, PSD2, open banking

1. Introduction

This Chapter describes advanced patterns relating to the use of cloud platforms in hosting IT solutions. The context for the patterns is the evolution towards open information ecosystems, mandated to a large degree by regulatory initiatives such as PSD2 [1], but also by competitive necessity. In this future business environment, there is an expectation of a significant increase in transaction volumes as new and innovative services become available for consumers.

The Chapter describes why cloud platform are the essential technology to provide cost effective scalability for IT solutions. The patterns highlight how new application components in the cloud are used in conjunction with existing on premises applications in a hybrid approach to deployment. Further, the Chapter also highlights how the patterns can then be used as part of a phased, but fully complete, migration to the cloud. Finally, specific real world usage scenarios for the patterns are highlighted.

The patterns are presented in a cloud vendor agnostic way and can be implemented in any of the key cloud provider technologies; Amazon Web Services (AWS), Google Cloud Platform (GCP) or Microsoft Azure.

The Chapter is structured as follows; Section 2 provides the background in terms of the business environment and the associated business drivers that necessitate the move to cloud. It further explores the technology perspectives of cloud that provide the specific advantages over conventional infrastructure technologies to support the emerging business environment. Section 3 introduces the definition of a cloud platform in the specific business context outlined.

Section 4 provides the underpinnings of the key cloud platform patterns in the form of relevant established architecture patterns and outlines the essential building blocks for cloud solutions. Section 5 then uses these underpinnings to describe the proposed cloud platform architecture patterns. Section 6 illustrates the use of the cloud patterns to achieve a migration from conventional IT solution deployment via a multi-phased approach. Finally, in Section 7, real-world scenarios are described to which the cloud patterns are directly applicable.

2. Background

Cloud computing has revolutionised the provisioning of infrastructure for IT services. As the maturity of the cloud offerings has increased, the richness of the of capability has progress from, initially, Infrastructure as a Service (IaaS), through to Platform as a Service (PaaS) and then Software as a Service (SAAS). The latest generation of cloud services relate to the availability of cloud platforms; domain specific applications connecting users of a particular services with the service providers via the concept of a cloud platform. There are numerous examples now appearing but, one of the earliest and a classic example of such is Uber.

The Chapter introduces some key business drivers for the use of cloud platforms. Specifically, the context of regulatory changes relating to open banking are used to illustrate trends in financial services domain. Its consequences in terms of impacts to IT system non-functional requirements, particularly those relating to the ability to scale on demand and cost effectively, are highlighted. This creates a problem for an organisation's IT function in supporting these trends.

The use of cloud technology and cloud platforms is now ubiquitous in most organisations IT architectural thinking, with the promise of providing:

- On demand and self-service characteristics; hence being suitable for agile delivery lifecycles
- Highly scalable architectures supporting platforms having huge global reach
- Capability to store of huge volumes of data and derive useful insights to inform a variety of downstream services

Various patterns for the migration of components to the cloud have been identified previously [2]. These have focussed on basic technology migration patterns such as:

- Re-Deployment
- Cloudification
- Relocation

This focus of this Chapter is in defining advanced, application migration patterns that exploit the advantages of cloud computing for use in emerging and future open information ecosystems. The patterns highlight the essential adoption and use of cloud platforms through:

- Enabling the caching and aggregation of customer data from which insights can be determined and further support downstream customer services

- Catering initially for a hybrid co-existence with 'on premises' IT systems and services such that these can be retained in the short term and provided cost effectively

- Ultimately supporting the complete migration of IT systems from on premises deployment to a cloud platform

2.1 Business drivers

The introduction of new financial market regulation, notably the revised Payments Services Directive [1] (or PSD2), has mandated banks to open up account information and payment services to third parties. The regulation is considered an important enabler for the creation of new and innovative customer propositions. PSD2 is recognised as a trigger for the wider concepts of 'open banking' and, beyond this, 'open finance' in which ultimately a rich variety of financial services are accessible to an ecosystem of third parties comprising third parties, business partners and industry bodies.

Open banking has led to a corresponding rise in financial technology organisations – namely the "*fintechs*". Indeed, information pertaining to the take up of open banking services confirms that 94% of fintechs view open banking as a major area of opportunity [3].

The net effect of this is that there is likely to be a growth in financial transactions accessing customer accounts as new services, founded on the open access regulation, are brought to market. This presents a challenge for financial institutions; that of scaling their IT systems cost effectively to support the new market dynamics with increased transaction volumes.

To summarise:

- Within financial services, open banking is recognised a key area for driving business growth

- To enable rapid pace of change and innovation, businesses must adopt technology that scales effectively and enables an engaging customer experience

- Cloud technology is considered essential to achieve scalability of the banks underlying systems to meet the demand for future services

Organisations are therefore faced with a key problem to address of how to combine the value of their existing mature core applications with the advantages that cloud technology provides. Superficially, they are faced with a number of high-level architecture challenges:

- Do they lift and shift applications to the cloud?

- Do they invest in rewriting applications to take advantage of the cloud technologies?

- Do they migrate to greenfield cloud platforms providing new implementations of core services?

This Chapter helps organisations to address these key issues by describing a variety of patterns highlighting relevant cloud platform usage scenarios including hybrid deployments and patterns to support, ultimately, the full migration of services to a cloud platform.

2.2 Cloud technology drivers

In this Section, a precis of the features of cloud technologies is provided. These reinforce what makes a cloud platform suitable for supporting the provision of scalable information services in general and the specific emerging trends in banking. The notable technology features are:

- **Elastic scaling.** As transaction volumes increase or decrease, cloud autoscaling technologies scale the computing resource required automatically.

- **Compact Data Notation using Java Script Object Notation (JSON).** Compact data representation standard based on name-value pairs. Again this infers requires less bandwidth than other data formats such as XML when transmitting data, making it more suitable for internet usage.

- **RESTful API standards**. A RESTful API uses standard HTTP requests to invoke remote processing on data. REST is the preferred API technology of choice as it is based on open standards and the use of a 'lightweight' stateless HTTP protocol for its requests and responses. This again reduces computing and networking resource requirements and makes a cloud solution inherently more scalable.

- **'No SQL' database technology**. 'Document' style storage databases allow for the storage of data is the same format as which it transmitted, specifically JSON. This approach requires no, or minimal, format translation from data store through to payload, further reducing the computing resources required in supporting a transaction from data deserialization through its transmission and presentation to a consumer.

- **Use of Open Source middleware**. Open Source middleware is prevalent for cloud deployments. Such software licencing models are much more scalable as the software is free, or at the very least the pricing models are better geared to the highly elastic solutions enabled by the cloud. Thus, cloud solution runtime costs are close to that of a linear 'utility' model, rather than having the incremental costs breaks associated with traditional middleware and vendor software. Hence this ultimately more cost efficient.

Consider now a traditional IT architecture that supports banking and other information systems. The underlying customer information will typically reside in a 'system of record', the implementation of which typically fall into one of two categories:

1. A bespoke legacy system, such as a mainframe, developed over many years; often difficult to maintain with rigid release cycles for enhancements

2. A vendor package, providing a complete or modular domain solution. e.g. a core banking platform or payment engine.

In the context of open information access, the ability to scale on demand and at a cost that is linear to the transaction load becomes a key requirement. However, both of the above implementation options present challenges in meeting the non-functional characteristics of a new open information ecosystem outlined since the ability to scale cost effectively becomes difficult.

One reason for this is that both legacy and vendor product pricing are typically very dependent on supporting hardware and the number of CPUs required. Hence cost breaks relating to hardware and vendor product licencing tend to be highly non-linear. To accommodate the open information ecosystem and expected transaction growth via a traditional IT architecture typically means over-engineering to allow for sufficient headroom in the capacity. Thus, mitigating the scalability risk in a traditional IT architecture is likely to be highly cost inefficient given the wide range of loads that could be experienced.

2.3 Open information ecosystem requirements

The difference in usage profiles between open banking and traditional banking are now explored. In general, traditional banking is subject to highly predictable loads based on:

- A finite customer base for a given banks
- Online usage patterns that are well understood and predictable
- System processing that is based on periodic cycles
    - Daily processing cycles such as overnight batch processing
    - Monthly processing cycles, such a billing

It is reasonable to assume that net transaction volumes will inevitably increase substantially as third parties develop their propositions and these gain maturity in the marketplace. This alone will result in customers interacting with their bank more frequently, albeit indirectly via the third parties applications in *'customer present'* scenarios. Also, there will be an increase in transaction volume driven from the third parties directly. Third parties, having obtained consent from the customer for specific account information, will exercise their right under PSD2 to access that information up to four times daily in *'customer not present'* scenarios.

However, with open banking, transactional loads are likely to be significantly less predictable. The open banking transaction volumes have a more complex and less deterministic relationship with existing customer volumes and their access patterns:

- Customers may employ the services of several third parties and thus a multiplier will apply to the volume of transactions normally associated with a given customer base. This multiplier is difficult to quantify as:
    1. The percentage of account holders that subscribe to use PSD2 services is not yet known
    2. The number of PSD2 services that customers subscribe to is likely to be highly variable
- third parties will undoubtedly access account information and transaction history without the customer being present up to the limit defined by the regulation.

- Information access patterns are less predictable and determined by the third party rather than via predictable customer access patterns that are well understood by the banks'.

These characteristics translate to specific IT issues for the account information provider, notably:

- How to achieve scalability of the mandated services to meet a, potentially huge, increase in transactions volume
- How to accommodate peak loads at non predictable times
- How to ensure performance and availability of the regulatory interface to support the open information services

3. Cloud platform approach

Given the challenges highlighted for open information access, the role of a cloud platforms in the providing solutions to this problem have previously been identified [4].

3.1 Platform definition

In brief, a platform is a business based on enabling value-creating interactions between external producers and consumers. The platform provides an open, participative infrastructure for these interactions and operates within governance conditions set for them. The platform's overarching purpose: to consummate matches among users and facilitate the exchange of goods, services, or social currency, thereby enabling value creation for all participants.

3.2 Technical service provider platforms

A Technical Service Provider (TSP) is a non-regulated participant in the PSD2 ecosystem. They provide services on behalf of a regulated entity and provide the necessary IT components to implement the required PSD2 services, intermediating between an Account Provider and a Third Party Provider via their platform, as illustrated in **Figure 1**. Standards for PSD2 access to account services (e.g. from the Berlin Group [5]) universally employ application programming interfaces (APIs), these being the de facto standard for B2B interfaces over the Internet. Further, as the ecosystem expands to accommodate broader open banking services, there is an expectation that additional, non-regulatory, services will also be implemented using APIs.

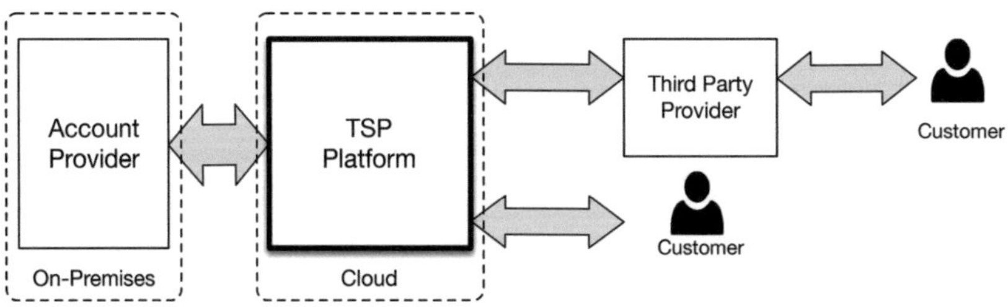

Figure 1.
Cloud platform context.

TSP platforms can accommodate such open banking services on behalf of an account provider.

3.3 Summary

The concept of a cloud platform has been introduced and the associated advantages highlighted. In practice such a platform can either be provided by a third party, known as a TSP, or by the bank themselves in the form of a private cloud. For the purposes of the patterns now described below and their rationale, this distinction is not significant.

4. Pattern building blocks

4.1 Command query response segregation

Command Query Response Segregation (CQRS) is a fundamental design pattern identified by Young [6] and Fowler [7]. Up until recently, the use of this pattern was quite limited and, furthermore, its usage came with caveats about the additional implementation complexity required. Through an implementation of this pattern,

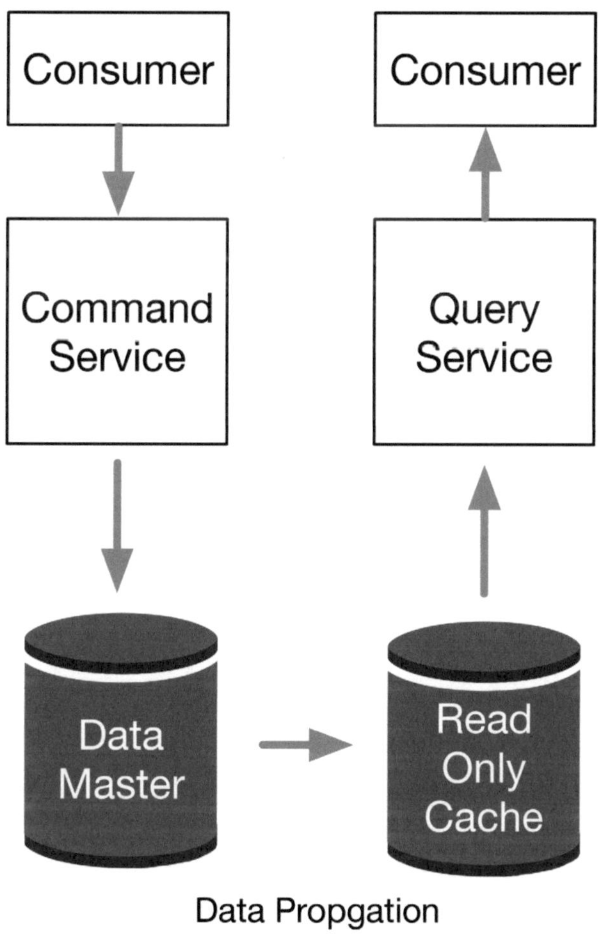

Figure 2.
Abstract CQRS pattern.

it will be shown that certain key benefits of a cloud platform can be realised. The pattern is shown conceptually in **Figure 2**.

In its abstract form, the pattern is very simple:

- One mechanism is used to read data – the *Query* element of the pattern
- A different mechanism is used to write data – the *Command* element of the pattern
- Data subject to an update in the *Data Master* (illustrated) is propagated to the *Read Only Cache* once an update has occurred occur.

The pattern is unspecific regarding implementation.

The significant feature of this pattern is that is reduces loading on the Data Master, since only write operations are performed on this data store. In the context of financial services this is highly significant. Consider the Data Master as supporting a system of record such as an accounting application. Since the majority of transactions on accounts are in fact read operations (typically 80%), by having a separate data cache for read only transactions, this approach becomes highly effective in reducing the net load on the system of record.

In the cloud patterns described in this Chapter, it will be shown how the query service and the command service can be implemented independently and deployed to a cloud platform in a phased approach if necessary. This enables scalability and performance and ultimately can facilitate a complete migration of a system to a cloud platform.

4.2 Publish-subscribe architecture

Publish-Subscribe is an architectural pattern that is exploited in the proposed patterns for cloud platforms. The components of the pattern are illustrated in **Figure 3**. The pattern is fundamentally about message distribution:

- An Event Publisher creates and sends a message
- An Event Subscriber receives and processes messages
- The messages delivery is facilitated by a Publish/Subscribe Engine.

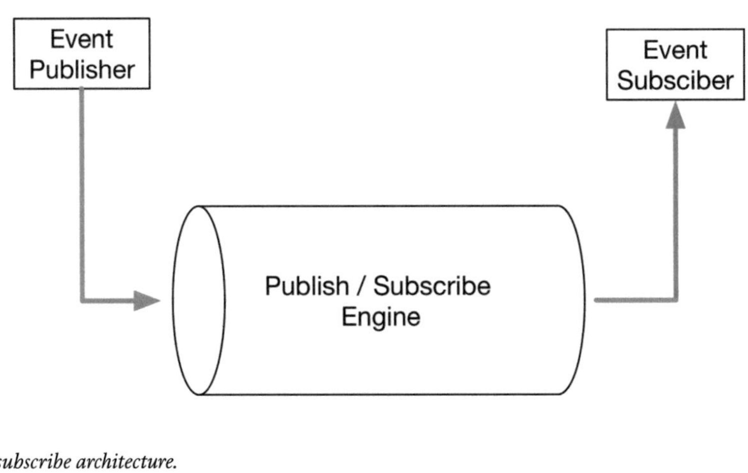

Figure 3.
Publish - subscribe architecture.

The Publish/Subscribe engine manages the distribution of messages based on the set of subscriptions. When an event is published, the engine matches the subscribers, typically based on the assignment of 'topics' and transports the message to the destination accordingly.

4.3 API gateway

A brief description of an API Gateway is provided here for the purpose solely of illustrating its role in the cloud patterns. In short, an API Gateway provides services for the management of APIs. These services broadly equate to a set of policy driven capabilities that dictate the characteristics and behaviour of an API. Typical policies relate to:

- Security management of the API
- Performance management, viz. throttling of endpoint connections

In addition, an API Gateway acts as an audit point and the logging of API usage. A number of commercial and open source API Gateway offerings are available.

4.4 Micro services architecture

The Service Oriented Architecture (SOA) paradigm has previously dominated architectural thinking. This paradigm relied on the constructs of a layered hierarchy of web services to fulfil a request. The services were specified and implemented via strongly typed interface definitions using XML. Similarly the invocation protocol, SOAP, was a verbose XML implementation.

Microservices have a similar concept of an interface definition but this is specified and implemented using a much simpler data typing language, namely JSON with services invocation via a 'lighter', stateless protocol, denoted Representational State Transfer (ReST).

However, whilst there are technical differences in the way that web services and microservices are specified and invoked, the difference in architectural style goes much deeper that the underlying technologies. Specifically, a microservice architecture has the following characteristics:

- It is not a 'layered' architecture in that each microservice should be designed to perform a specific function through from data presentation to the data persistence
- Each microservice should therefore encapsulate all the functionality to support:
    - Presentation of data, irrespective of whether presentation layer is a graphical or 'headless' data payload.
    - Business logic associated with the service. e.g. business validation logic.
    - Data retrieval and update services.
- They employ a 'lightweight' ReST protocol for the invocation of each microservice.

It is useful to emphasise the difference between this and the traditional service-oriented architecture paradigm as this is key to the effectiveness of the cloud platform patterns presented here. In order to compare, **Figure 4** illustrates the typical layering of a SOA architecture. This is shown alongside the concept of a microservices architecture, with each microservice encapsulating a vertical 'slice' through this layering. In its simplest form, the microservice architecture is a series of such vertical slices with each microservice being a completely independent construct and having zero coupling to other microservices.

4.5 Event based architecture

An event-based architecture is another mature architecture paradigm that complements perfectly a microservices architecture by supporting the communications between them. The design principle for this style of communication, again relates to ensuring decoupling between microservices. Independence of each microservice supports the ability to design, build and deploy microservices without impact to other microservices supporting the concept of Domain Driven Design [8].

Thus, rather than create dependencies between microservices using point to point connections between them (i.e. one microservice explicitly invoking another microservices via it interface), using an event driven architecture pattern, a microservice will publish data via an event construct. Microservices that are potentially impacted by the event, subscribe to the event and receive the event and its associated data.

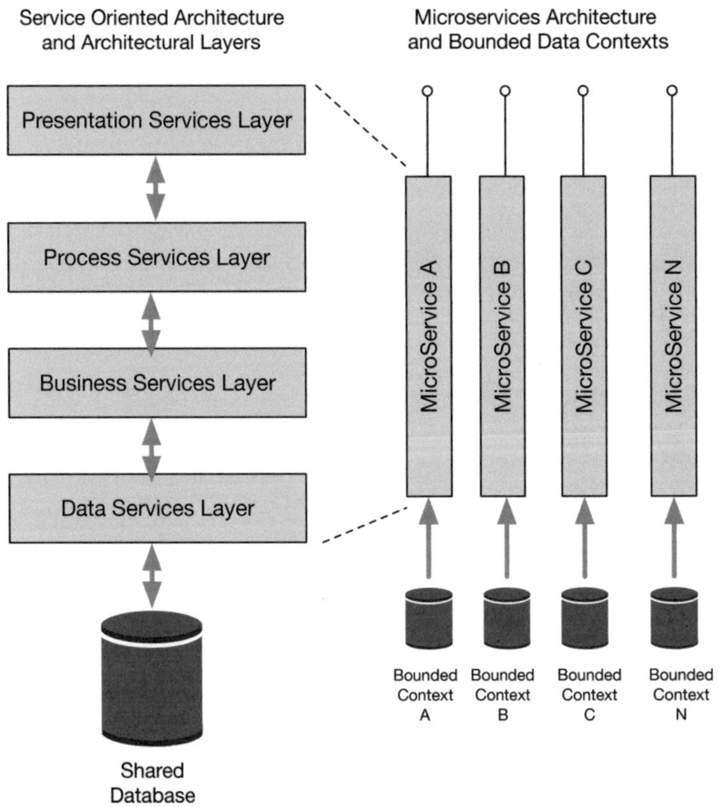

Figure 4.
SOA versus microservice architecture paradigm.

To implement this architecture pattern a Publish-Subscribe Engine component is required. This component manages the publication of events, typically via the definition of topics. Consumers of the events are defined by their subscriptions to the variety of topics.

4.6 Bounded data context

In its general from, a bounded data context defines the necessary data entities and attributes to support a given business domain. This is another idea that is integral to the concept of Domain Driven Design [8]. In simple terms, a bounded data context defines a key domain entity such as a Customer, an Order or an Account. The data attributes for the bounded data context contains foreign keys that allow linkages to other domains. For example, a 'Customer' bounded context may contain a list or collection of 'Account IDs' to define linkages to the Account domain entities that equate to the Account bounded context. Limited hierarchical nesting of the data enables implementation using JSON data type definitions rather than a stronger typed data structure implementation such as XML.

By constraining the data set in this way, a relatively simple data structure can thus support the microservice in respect of its CRUD services. Since the bounded contexts each define a self-supporting and independent data set, this in turn supports low coupling between the microservices allowing for independence of design, through to packaging and deployment.

4.7 No SQL databases

SQL databases relate to a very specific data entity relationship model and associated query model based on tables. NoSQL databases are a technology that provide an alternative to traditional data entity relationship models and storage and support data retrieval mechanism that are different to the traditional entity relationship model. There are a several fundamentally different types: columnar, document (aka object) databases, key-value pair and graph NoSQL databases.

The bounded data context approach is well suited to an implementation using a NoSQL database, specifically the document style. Data can be serialised and de-serialised efficiently without any paradigm shift in the data representation. In this respect, JSON microservice payloads can translate directly to serialised document objects and vice versa.

5. Platform patterns

A prime focus of the patterns for cloud platforms presented in this Chapter is the notion that you for a specific functional service, you use a different approach to update information than the approach you use to read information. The original idea stems from a pattern known as Command Query Responsibility Segregation outlined in Section 6 above.

Consider now the business context attributed to open banking described in Section 2. When deconstructed into its constituent parts, the CQRS the pattern can be seen to be highly useful in supporting organisations in meeting a number of the business drivers that have been identified, specifically:

- To meet their regulatory requirements for open access to their customer's data

- To scale ageing legacy systems cost effectively to meet growth needs
- To ultimately support migration of complete legacy platforms to a new cloud-based platform, helping to meeting the requirements for an agile IT organisation supporting IT changes with a high cadence.

In this respect this Section highlights the following patterns each using elements of the original CQRS pattern to meet a specific use case. The following cloud platform patterns are identified:

- Cloud Data Cache
- API Façade
- Data Hydration

Further, it is shown that, through the sequencing of these new patterns, they can be used to facilitate the complete migration of a on premises legacy system of record to a new cloud platform. The cloud patterns are now described in detail.

5.1 Data cache

This pattern relates to the provision of read only services via a cloud platform. The business context of this pattern is that information services, traditionally provided by an organisations' core systems, such as a system of record, can now be invoked indirectly via a third party, such as is mandated by the open banking regulation. In such a scenario there arises an increase in demand for services. This in turn places demands of increased transactions and additional load on the core system of record.

Consider now that, for applications such as core banking systems, read transactions typically account for 80% of transaction volume. In these circumstances, to alleviate transaction load on the core system, the pattern provides a read-only data cache of data derived from the system of record. As described in Section 2.2, this solution therefore provides a highly scalable solution to this particular business scenario by using the cloud platform pattern, notably through:

- Use of low-cost, open-source licencing models for the cloud component technologies and middleware.
- Avoidance of high cost, monolithic scaling of the underlying system of record having high cost-breaks.

The key advantage of this pattern is that, in response to such increased demand for read only information services, the organisations information services can be scaled in a far more cost-effective way than by scaling the underlying system of record.

The components of this pattern are illustrated in **Figure 5** and their role described in **Table 1**.

5.2 Hydration engine

This pattern relates to population of the data stores cache to support the Data Cache Pattern described above. Each of the populated data stores represents a bounded data context for the microservices in the Data Cache pattern above.

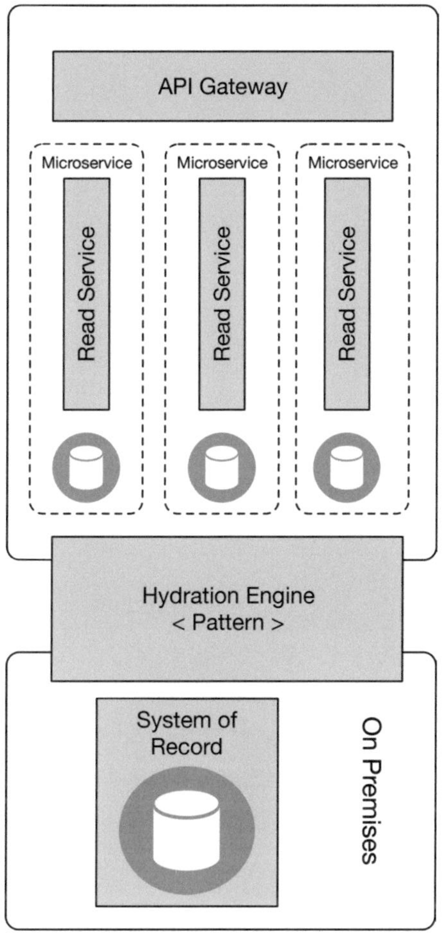

Figure 5.
Data cache cloud pattern.

Component	Role
API Gateway	Hosts proxy APIs for each microservice and acts as the Policy Enforcement Point (PEP) for access to the services.
Microservice	A microservice is associated with each API offered via the Gateway and provides read access to a given data cache.
Data Cache	A number of data stores are provided each relating to a single bounded data context
Hydration Engine	Its purpose is to populate the data caches and keep them synchronised with the system of record. This component is itself a pattern and may have a number of implementation variants.

Table 1.
Data cache pattern components and role.

The population of the data stores relies on the implementation of the publish-subscribe pattern described in Section 4.2 To populate the data stores, a single subscriber microservice is defined for each bounded data context identified.

The components of this pattern are illustrated in **Figure 6** and their role described in **Table 2**.

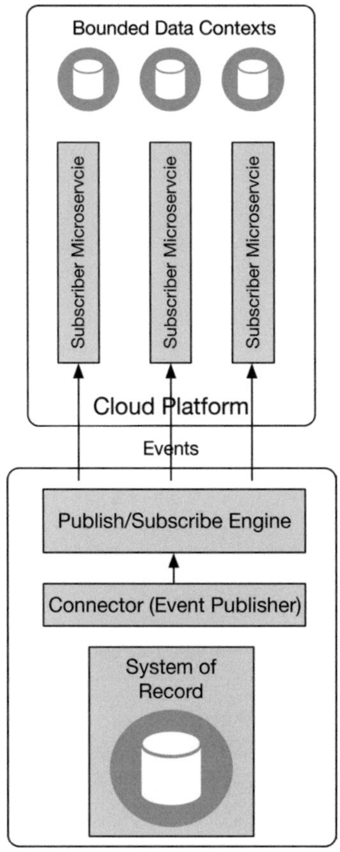

Figure 6.
Hydration engine cloud pattern.

Component	Role
Data Cache	A number of data stores are provided each relating to a bounded data context.
Subscriber Microservice	Each subscriber microservice subscribes to a set of event services sufficient to populate the bounded data context of the data store.
Publish/Subscribe Engine	This component maintains the set of events for publishing data and maintains the set of subscribers to the events.
Connector	The connector provides integration with the source system. The connector detects changes in the underlying data in the system of record and translates these into events for processing by the Publish/Subscribe Engine.
System of Record	This component represents the master application that is the source of the data being cached.

Table 2.
Hydration engine pattern components and roles.

5.3 API Façade

This pattern is a cloud specific implementation of the well-known Bridge pattern [9]. The pattern assumes that there are an existing set of services that provide integration with the system of record. The pattern implementation provisions a set of modern API interfaces, functionally equivalent to those provided by the combination of the system of record overlayed by its existing integration services. Such existing integration services could be implemented by a number of technologies, including:

- SOAP web services
- Messaging services, such as MQ Series, Rabbit MQ, AWS Simple Queuing System
- CICS transaction processing technology

In this respect, the APIs represent a new interface definition and constitute a 'façade' for the existing integration services. The APIs are hosted within the cloud platform, fronted by an API Gateway that provides API management controlled through policies as described in Section 4.3.

The components of this pattern are illustrated in **Figure** 7 and their role described in **Table 3**.

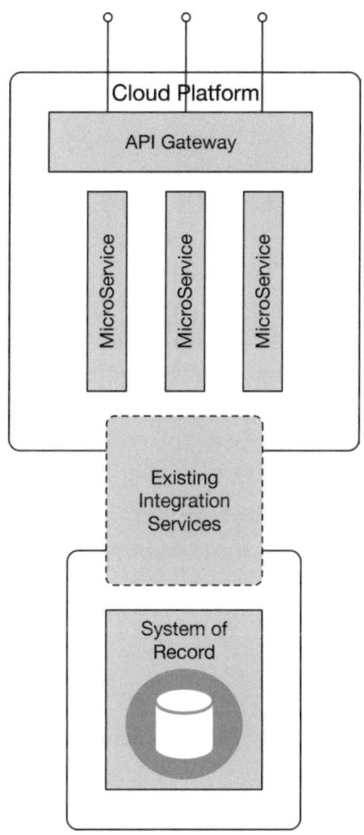

Figure 7.
API facade cloud pattern.

Component	Role
API Gateway	Providing policy based access to the APIs.
Facade Microservice	Provides the implementation of the API interface by consuming the existing legacy integration services.
Existing Integration Services	This component represents existing integration services that are consumed by the new microservices to access the underlying systems of record.
System of Record	This component represents the master application that is the source of the data and the target of data updates.

Table 3.
API facade pattern components and role.

6. Cloud platform migration

This Section provides an illustration of how the three patterns outlined previously can be used to achieve a migration of an on-premises legacy application, such as a system of record, to a modern cloud platform. Three potential phases of the migration are identified, fulfilling gradual 'strangulation' [10] of the legacy platform as shown in **Figure 8(a)-(c)**.

Phase 1 of the migration provides a set of selected read services via the cloud platform using the *Data Cache* and the *Hydration Engine* patterns. This migration step itself can be a phased approach, gradually incrementing the number of the bounded contexts that are supported in the cloud platform.

Phase 2 of the migration then provides complementary write services for the read services and their bounded data contexts. To affect this migration the *API Façade Pattern* is used to support the write services. Having both read and write services for a given bounded data context allows integration in the form of update via events between the system of record and the cloud platform to be switched off.

A caveat to this happening is that the consuming applications must be migrated to use the new cloud platform services and not continue their use of the legacy services. Without this occurring, the architecture becomes complicated by the fact that any updates made via the cloud platform must also be propagated back to system of record. To support this, the system of record must also be a subscriber to events derived from updates via the write microservices. Similarly, any updates made to the system of record must be propagated to the cloud platform. To support the latter, the hydration engine must be retained in the architectural solution at this stage.

To avoid such a complication requires the coordination of:

- Provision of the write services to complement the read services within the cloud platform for each of the bounded contexts.
- Migration of the service consumers to the new cloud platform services as they become available.
- Discontinuing use of the equivalent legacy services.

Achieving a clean separation of write services can be difficult, particularly if there is not a simple correspondence between the legacy and cloud platform services. Similarly, Phase 2 can represent the target state architecture where a residual set of legacy services are retained that existing consumers still continue to use. This is very much a realistic scenario in the cases where it not feasible to change the legacy consuming applications to use the new API based cloud services. A strategy of maintaining the legacy system of record for legacy consuming clients may therefore be necessary. Alternatively, new consumer applications may be built to complement the existing legacy client applications and the legacy clients may ultimately be deprecated.

If the intention is to fully deprecate the system of record, then the migration process can proceed on a per bounded context basis until all the data that was originally managed by the system of record is represented in the cloud platform. The consumers of the legacy services must be migrated to consume the new services offered by the cloud platform as the bounded contexts are gradually

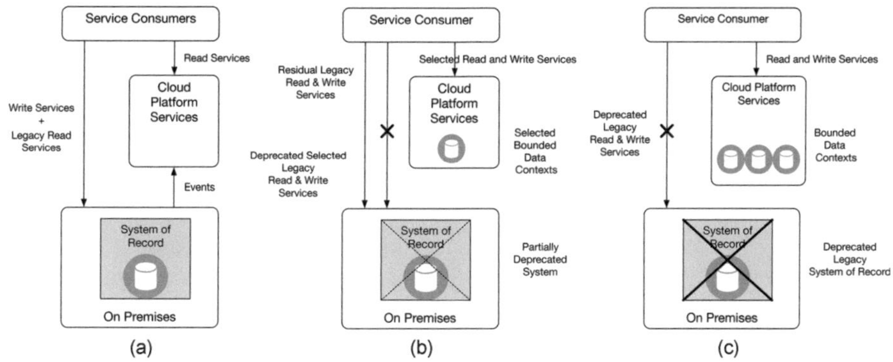

Figure 8.
(a)-(c). Migration to a Cloud Platform via Strangulation.

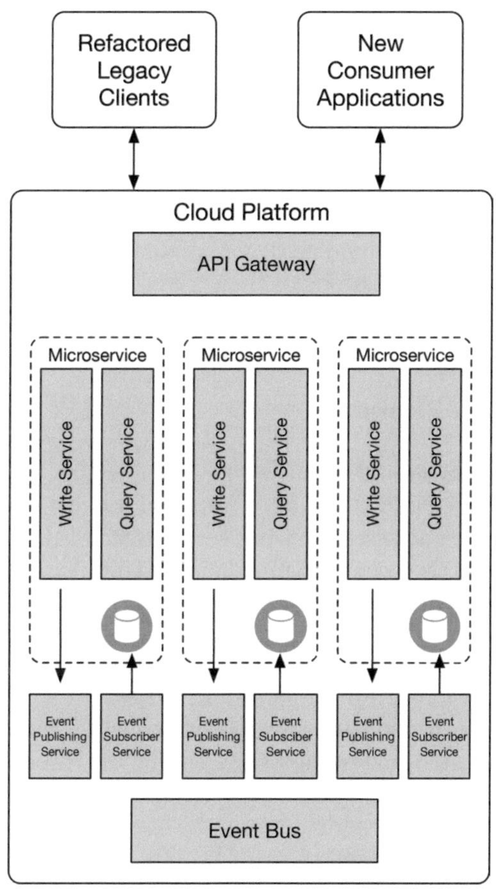

Figure 9.
End state architecture post migration.

migrated. Once the migration has completed, the legacy system of record can then be deprecated. In these circumstances, to replace the legacy client application, refactored or completely new client applications must be built on top of the new cloud platform services. The end state architecture in post migration is shown in **Figure 9**.

7. Pattern usage scenarios

This Section describes four specific example usage scenarios for the cloud patterns introduced here.

7.1 Open banking account and transaction data

Regulatory initiatives such as the PSD2 [1] in Europe and the Competition and Market Authority Order [11] in the UK dictated banks must provide information services relating to customer's account details and their historical transactions to approved third parties, subject to customer consent.

Using the data they obtain about the customer from their banks, the third parties are thus empowered to create innovative, value add, services that entice the bank's customers. These new services create a demand profile for information from the banks that is significantly different to existing customer behaviours; these being typically highly predictable and with a tendency to be based on a point in time transactional need e.g., to check their account balance, to make a transfer. Given that third parties are permitted to access customer information multiple times per day, this will result in a significant increase in transaction frequency from the banks perspective as third parties will take advantage of this to keep their data up to date to reflect a given customer's intra-day transactions.

As explained in Section 2.2, faced with a choice of scaling their existing accounting systems of record to accommodate this increased transaction volume, the bank should implement the Data Cache Pattern described in Section 5.1 and the supporting pattern in Section 5.2 to achieve cost effective scaling to support the increased transaction volumes.

7.2 Public and private API hosting

Once again, in response to the landmark regulatory initiatives for open information access previously described, financial institutions, notably banks, are mandated to provide access to account services to third parties. In terms of the technology to offer these services the de facto architectural style for implementation of these services is that of ReSTful APIs. Similarly, to complement the regulatory services, banks may also choose to offer their own services for consumption by their partner organisations or to monetise additional, non-regulatory services for consumption by third parties.

The net effect of this is that banks need to offer a wide range of ReSTful API services for consumption by external parties. To support these services bank should implement the API Façade cloud pattern of Section 5.3, enabling controlled policy based access to the set of APIs implementing the functional services and leveraging existing integrations to the systems of record where appropriate.

7.3 Customer data aggregation

A third usage scenario relates to the aggregation of customer data. Third parties access and accrue account information for a given customer from multiple financial institutions. This data should be captured according to the Data Cache pattern and serves to support the third party provider in obtaining a convenient and full picture of the customer's financial position. This data supports their provision of value add services to their customers. A key observation is that the cloud platform implementation is provided by the third party provider, not by the account provider, resulting in a demand side cloud platform [5].

A variant of this is that, within a given financial institution, data about a customer may be aggregated from a number of different account systems of record (e.g., current account savings account, credit card account) via the same Data Cache Cloud Pattern. Conversely, this usage scenario represents a supply side platform. The two styles of platform are illustrated in **Figure 10** below.

7.4 Legacy system remediation

A common problem for banks and other financial institutions is that of vendor lock-in to legacy technologies caused by a variety of circumstances:

- Low risk appetite of the organisation to undertake a complex migration to a new replacement system of record

- Sheer effort to refactor the legacy application using a modern IT architecture

At the same time, drivers to move from the legacy platform have increasing immediacy:

- Scarcity of resources to maintain and enhance the legacy system

- Correspondingly high maintenance costs

- Inability to support business resiliency due to slow development timescales and long delivery cycles for changes

In this context, the phased migration using the strangulation pattern outlined in Section 6 offers a viable solution to the vendor lock-in problem. By allowing for a phased migration the approach this significantly de-risks the migration to a new system of record.

- New services are introduced in a controlled manner, rather than one 'big bang'

- The approach has low initial complexity, focusing on read services for new consumers

- It has the advantage that legacy application service consumers are not initially impacted by introduction of new services.

8. Summary

This Chapter has highlighted the key business and technical drivers to leverage cloud platforms in an era of open information services. Specific examples and scenarios from the financial services domain have been provided, but these are considered readily able to generalise to other business domains.

As open access to information becomes more prevalent, either though regulation or competitive necessity, there will be a need to support increased volume of transactions to access information. In these circumstances, to support scalability of the underlying information systems, it is considered vital to leverage the properties of cloud infrastructure. To do this effectively the key architecture patterns have been identified to support this business prerogative.

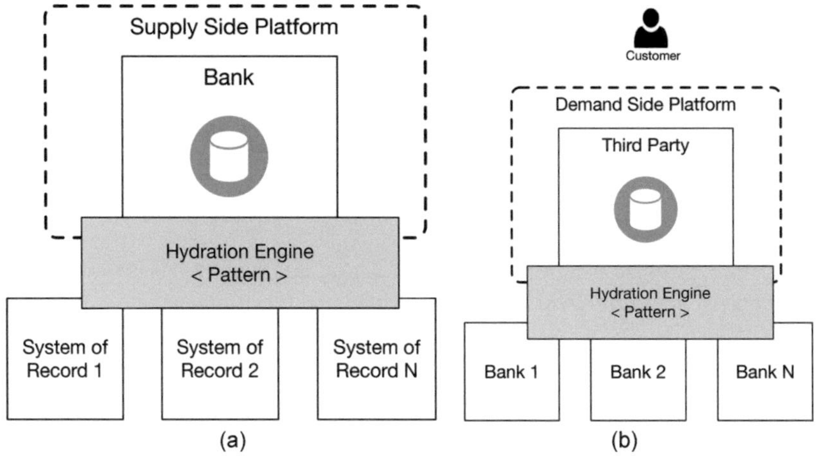

Figure 10.
(a) Supply side and (b) demand side cloud platforms.

The patterns accommodate both:

- A hybrid approach, leveraging existing infrastructure, co-existing with a cloud platform and;
- A phased, but ultimately complete, migration from a conventional infrastructure deployment to that of a cloud platform

The patterns have been presented in a cloud provider agnostic manner and there are a significant number of technology implementations that can be considered that map to the middleware capabilities that have been highlighted. This makes them highly realisable with current cloud middleware technologies and the key global cloud providers; AWS, GCP and Azure.

Glossary

API	Application Programming Interface
AWS	Amazon Web Services
B2B	Business to Business
CQRS	Command Query Response Segregation
CICS	Customer Information Control System
CRUD	Create Read Update Delete
GCP	Google Cloud Platform
HTTP	Hypertext Transfer Protocol
IaaS	Infrastructure as a Service
JSON	Java Object Notation
MQ	Message Queue
PaaS	Platform as a Service
PEP	Policy Enforcement Point
PSD2	Payment Services Directive 2
ReST	Representational State Transfer
SaaS	Software as a Service
SOA	Service Oriented Architecture

SOAP	Simple Object Access Protocol
SQL	Structured Query Language
TSP	Technical Service Provider
XML	eXtensible Markup Language

Author details

Gary S.D. Farrow
Triari Consulting Ltd, Manchester, UK

*Address all correspondence to: gary.farrow@triari.co.uk

References

[1] Directive (EU) 2015/2366 of the European Parliament and of the Council of 25 November 2015 on payment services in the internal market, amending Directives 2002/65/EC, 2009/110/EC and 2013/36/EU and Regulation (EU) No 1093/2010, and repealing Directive 2007/64/EC (Text with EEA relevance), available at: https://eur-lex.europa.eu/legal-content/EN/TXT/?uri=CELEX%3A32015L2366, last accessed on 7 Feb 2020

[2] *"Pattern-based Multi-Cloud Architecture Migration"*, Pooyan Jamshidi1, Claus Pahl2, Nabor C. Mendonc , Software Practice and. Experience. 2016; 00:1-25

[3] Ernst & Young (2018) 'FinTech Open Banking Snapshot', available at: https://assets.ey.com/content/ dam/ey-sites/ey-com/en_gl/topics/banking-and-capital-markets/ey-FinTech-open-banking-snapshot. pdf Last accessed 6 June, 2021.

[4] The Berlin Group (2020) 'NextGenPSD2 XS2A Framework — Implementation Guidelines, Version 1.3.6', available at: https://www.berlin-group.org/nextgenpsd2-downloads. Last accessed 8 Jun 2021

[5] *"Open Banking: The Rise of the Cloud Platform"*, G. S. D Farrow, Journal of Payments Strategy & Systems, Volume 14 Number 2, 2020.

[6] *Young, Greg. "CQRS Documents" (PDF)* http://cqrs.files.wordpress.com/2010/11/cqrs_documents.pdf. *Last accessed 7 May 2021*

[7] *Fowler, Martin. "CQRS".* https://martinfowler.com/bliki/CQRS.html. *Last accessed on 7 May 2021*

[8] *Domain Driven Design: Tackling Complexity in the Heart of Software. Eric Evans, Sept 2003, ISBN-10032112515*

[9] *Erich Gamma, Richard Helm, Ralph Johnson, John Vlissides (1994). Design Patterns: Elements of Reusable Object-Oriented Software. Addison Wesley. pp. 151ff. ISBN 0-201-63361-2.*

[10] An Agile Approach to a Legacy System, Chris Stevenson and Andy Pols, http://cdn.pols.co.uk/papers/agile-approach-to-legacy-systems.pdf. Last accessed 10 May, 2021

[11] 'Competition and Marketing Authority Report - RETAIL BANKING MARKET INVESTIGATION, "The Retail Banking Market Investigation Order"', 2017.

Chapter 5

Cloud Security in Middleware Architecture

Jagdish Chandra Patni

Abstract

The new Internet of Things (IoT) has increased the need for computing, connectivity, and storage capacities as the amount of sensitive data grows. Since it provides on-demand access to a common pool of resources such as processors, storage, software, and services, cloud computing can seem to be a convenient solution. However, there is a cost, as excessive communications burden not only the core network, but also the cloud data centre. As a result, it's critical to consider appropriate approaches and security middleware solutions. In this chapter, we define a middleware architecture to address security concerns and explore the general concept of cloud to achieve a higher level of security. Since it is designed to pre-process data at the network's edge, this security middleware functions as a smart gateway. Data can be processed and stored locally on fog or sent to the cloud for further processing, depending on the information obtained. Furthermore, the devices communicate via middleware, which gives them access to more computing power and improved security capabilities, allowing them to conduct safe communications. We discuss these concepts in detail, and explain how this is effective to cope with some of the most relevant security challenges.

Keywords: Internet of things, Cloud middleware, Software-as-a-Service, Platform-as-a-Service, Infrastructure-as-a-service, Amazon web service, Microsoft Azure, Virtual machine, Virtualization

1. Introduction

Cloud security ensures the secure cloud computing environment from both the internal and external cybersecurity attacks. Cloud computing, deliver the services to the end users by using information technology tools and methods that is now most demanding area of research for the public as well as private sectors those want to accelerate in the field on research and innovation. Widespread use of cloud computing technology also emerge the security challenges to the cloud developers. It becomes more interesting to create the cloud security solutions to prevent from unauthorised users or cybersecurity attacks/threats [1].

1.1 Cloud computing categories

Basically four main categories of cloud computing in practice and they are as follows:

1.1.1 Public cloud services, by public cloud provider

Public cloud services are the services provided by the public cloud providers, are SaaS, IaaS, and PaaS.

In public cloud type all the computing resources are available for the public use via internet [2]. All the resources may be varied depending upon the services providers but all include storage, applications or virtual machines. It provides the resource sharing and processing power distribution that is difficult to achieve by an organisation on its individual capacity.

Some public cloud services are free to use for all the users and some services are restricted to selected individuals or organisations. The use of resources are available to rest of users by paying the charges or subscriptions that vary one to another. That will save the huge amount of money of an organisation that want higher processing speed without setting its own setup (**Figure 1**).

While cloud services are used by public, security becomes the major concern and that need to be addressed properly. To address the security concern we needed the experienced staff and set of methods and protocols those can deal with security. Strict policies and procedures have to deployed to protect data from other different intended users.

1.1.2 Benefits and challenges of public cloud

The cloud services provides the faster and complete solution that is really not possible with the individual capacity of an organisation. It also ensures that no need to go for additional hardware and software solutions once the business is growing.

Cloud based services and applications can be used with the help of less hardware and software with great performance. We can also explain that end users not need to worry about installing and updating the hardware as well as software. It always ensures that all the applications will be up to the mark all the time without investing too much infrastructure and budget.

Public cloud helps to the organisations to grow without accumulating substantial costs. Examples of public cloud include like Amazon AWS, MS Azure are charging as per the usage by customers or organisations that reduce the operational cost of the organisations.

Figure 1.
Cloud computing architecture [3].

1.1.3 Private cloud services, managed by a public cloud provider

The for the individual customers that can be operated by third party.

- Operated by internal staff — The cloud services are managed and maintained by data centre internal staff as per the individual customer requirement in a virtual environment they control.
- The computing will be dedicated to the individual business of individual entity. This can be setup by cloud providers on the customer premises.
- By deploying cloud services it enhanced the control given to the individual business organisation.
- It is a type of virtual private cloud that can be set as at the customer unit or by virtual environment (**Figure 2**).

1.1.4 Benefits of private cloud

The following are the benefits to use the private cloud [4]:

Security and compliance: Compliance is critical for companies operating in highly regulated industries. Since confidential data is stored on hardware that no one else can access, private cloud storage allows businesses to comply with strict regulations. This benefit is available both in on-premises hardware installations and in hosted services.

Customization: An on-site cloud architect builds a completely private cloud, allowing stakeholders to decide exact environment required to run specific applications. The benefits of private clouds are similar to those of on-premise private

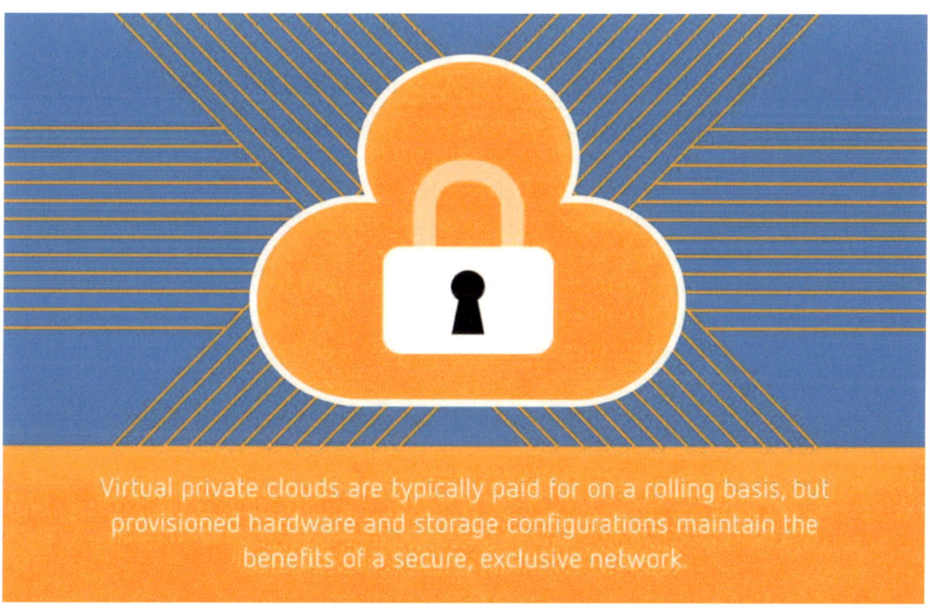

Figure 2.
Virtual private cloud [3].

clouds, but they do not need any on-site configuration. The company collaborates with a provider to set up and maintain a cloud that is solely for its use.

Hybridization: Hybridization expands the capabilities of a both private cloud and public cloud to ensure uptime without the need to mount new physical servers when an application requires more computing resources. This will be a cost-effective option for businesses that need protection of a private cloud also require the powerful services of public cloud for other functions.

1.1.5 Challenges of private cloud

If a company's computing needs aren't predictable, a private cloud can be problematic. When resource demand fluctuates, a private cloud can be unable to scale efficiently, thereby costing the company high investment. Some key considerations for IT peoples to think about.

Direct cost on Investment: It require significant investment to deploy the fully functional private clouds that hosted on-site and it may be value for the organisation after long time. Hardware cost is very high to establish a private cloud, and the environment would need to be set up, maintained, and managed by an expert cloud architect. Hosted private clouds, on the other hand, will significantly reduce these costs.

Capacity utilisation: The company is solely responsible for optimising resources usage in private cloud model. A cloud deployment that is underutilised can cost a company a lot of money.

Scalability: Where more computing power needed from private clouds, scaling the resources of private cloud that can take more time and money. This procedure would typically take more time to scale a virtual machine or requiring more services from a public cloud service provider.

1.1.6 Private cloud providers

Organisations who want to use the private cloud but do not have the funds to invest in an on-site solution will partner by using private cloud service provider. Some of the most well-known names in this field are:

Hewlett Packard Enterprise

HPE is a big name in the field of cloud computing era. Offering robust services as per the organisational needs. Customers can select hardware as well as software as per their needs.

Cisco

On-demand storage, advanced application performance management and automated container management are all available from Cisco. Data protection that have sufficient workloads to improve compliance is provided by Cisco solutions. Cisco have teamed up to provide stable application, desktop, networking and cloud delivery solutions to help businesses grow into digital enterprises.

Microsoft

Any corporate data centre will benefit from Microsoft's Azure Stack solution, which brings the power of an integrated cloud to any data centre. Azure is ready for hybridization, so businesses can take advantage of compliance features while still taking advantage of the full Azure cloud solution as required. Learn how Citrix and Microsoft are working together in the cloud to help you keep up with the pace of business.

Dell, IBM, VMware, Oracle and Red Hat are other big names in the field of private cloud providers [3].

1.1.7 Hybrid cloud services

Price, protection, operations, and access can all be optimised by combining private and public cloud computing configurations to host workloads and data. Internal personnel and, if desired, the public cloud provider would be involved in the operation.

A hybrid cloud services are combining the on premises computing infrastructure with private cloud services and public cloud services to get the higher computing, storage and services.

1.1.8 Hybrid cloud benefits

While cloud providers can help you save money, their real value comes from their ability to support a fast-paced digital business transformation. There are two priorities in any technology management organisation: various key points like IT and business transformation. Traditionally, The IT key points more focused on cost-cutting. ON the other hand digital transformation focused on making money from investments.

The main advantage of a hybrid cloud is its flexibility. A central concept of a digital company is the need to adapt and change direction quickly. To achieve the high performance and robustness the organisations combines all three public clouds, private clouds with on-premises resources [5].

1.1.9 What about hybrid Cloud good of bad?

Everything cannot belongs in the public cloud, the progressing businesses are opting for a hybrid cloud solution. Hybrid clouds combine the advantages of both by using existing data centre architecture.

This approach enables its components and other applications to communicate across boundaries, instances of cloud, and architectures. Data needs the same degree of delivery and access flexibility. In the complex digital world, whether you are dealing with workloads or databases, you can prepare for things to change around in response to changing needs.

1.1.10 Hybrid cloud scenarios

- **Dynamic workload Conditions**- For our dynamic workloads, use scalable public cloud and computing sensitive workloads on private clouds or in our private data centres.

- **Categories between critical and less-sensitive workloads**- We use a public cloud to compute our other business applications and other sensitive or critical applications on our private cloud.

- **Processing huge amount of Data**- It's unlikely that you'll be able to process large amounts of data at a near-constant rate. Instead, we could use highly scalable public cloud tools for our big data analytics and to keep our confidential data with complete protection we can use the private clouds with higher set of security systems.

- **Easy switching of data -** Use a public cloud or a private cloud for rest of the miscellaneous workloads. Also see the best suit for the organisation and switch accordingly between public and private.

- **Temporary arrangements of Resources**- Whenever our requirements are for a short time so instead to setting our cloud setup go with a public cloud that reduces the extra burden on us.

1.2 Comparison between public, private and hybrid cloud

A private cloud, also known as a corporate cloud, is either provided by a service provider or built on-site at a company's data centre. In either case, since the services are earmarked for particular users only, the private cloud appears to provide more protection.

As resource demand increases, a hybrid cloud environment extends a stable private cloud to a public cloud. This model enables businesses to remain compliant while still taking advantage of public resources. Organisations that use hybrid cloud will get the most out of their internal resources without causing a resource overload if demand spikes unexpectedly [6].

To access the applications and services from online computer is a key benefit of public cloud services. We can use the critical or complex applications virtually because the computer performs little to no computation.

To ensure smooth and fast disaster recovery, a service provider can store replicated files across multiple data centres. Public cloud platform also ensure the data safely from outside world that considered secure from the majority of threats. Public clouds can be configured differently:

1.2.1 Software as a service (SaaS)

In which a provider distributes its computing hosted in the cloud is known as software as a service (SaaS). The programme is accessed via the internet. Individual users are not required to install software on their personal computers under this model. This lowers the organisation's hardware requirements while also lowering service and repair costs.

1.2.2 Platform as a service (PaaS)

Platform as a service (PaaS) is a computing model that enables a company to build software without having to worry about the infrastructure. In essence, a provider creates and maintains an optimised environment that users can easily access by internet. Version control and compile facilities, as well as computing and storage tools, are often included in PaaS.

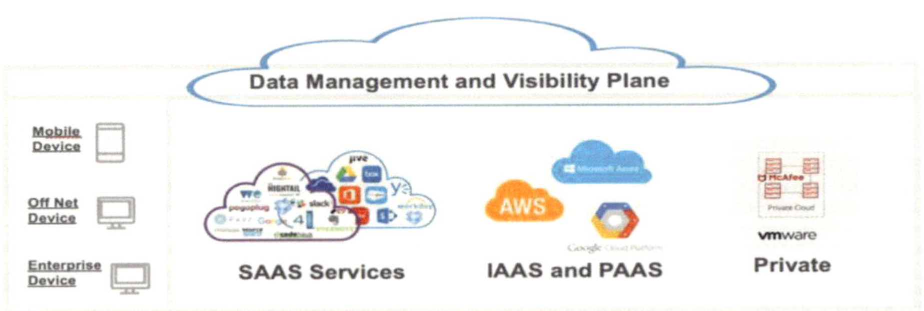

Figure 3.
Cloud services [2].

1.2.3 Infrastructure as a service (IaaS)

Infrastructure as a service (IaaS) is a business model under which a company outsources the entire data centre to a cloud provider. The provider manages the virtualization of the environment and hosts. Cloud adoption is made easier with IaaS. Purchasing and repairing hardware on-site is also more expensive than using the device (**Figure 3**).

2. Cloud security responsibilities

All the services like data, applications are hosted by a third party while using a public cloud computing service, which is a significant difference between cloud computing and traditional IT services, where data is stored within a network that managed by self. First and main step in developing a cloud security plan is to understand security responsibilities [7].

Major cloud providers strive to provide consumers with a stable cloud. Preventing breaches and retaining public and consumer confidence is central to their business model. Cloud providers may try to prevent cloud security problems with their services, but they have no control about how consumers use them, add data and those are going to access the data. The provider and the cloud client share various levels of security obligation in each public cloud service form. These are the different types of services:

- **Software-as-a-service**- Users are self-responsible to secure the data and its access.

- **Platform-as-a-service**-Users are self-responsible to secure the data and its access and applications used by them.

- **Infrastructure-as-a-service**-Users are self-responsible to secure the data and its access, applications used by them, operating systems and network traffic.

2.1 On-premise SaaS vs. PaaS vs. IaaS

Clouds were once all white fluffy stuff in the sky, and the IT services are restored at on-premise. Almost all of the applications and processes can now be run on the Cloud platform.

- **IaaS:** Cloud services as per use and pay option for various services like storage and virtualization.

- **PaaS:** Various tools in place of hardware and software that are available with cloud providers.

- **SaaS:** software's that we can use with the help of third party using cloud platform.

- **On-premise**: The software and services are going to be installed within the organisation (**Figure 4**).

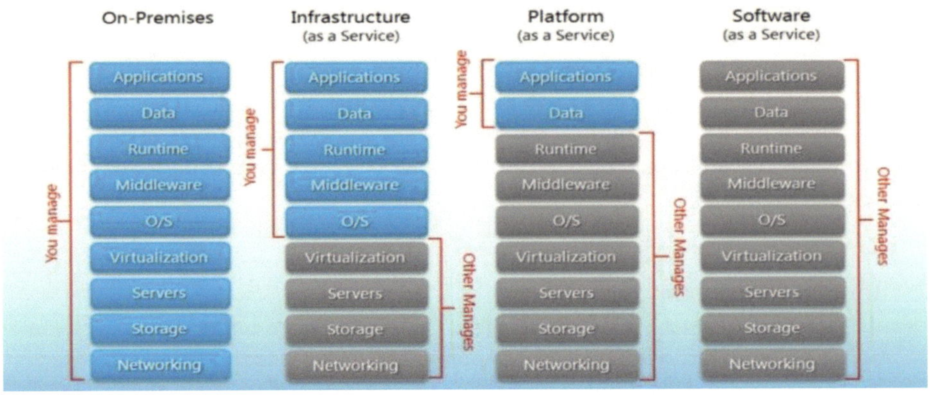

Figure 4.
SaaS, PaaS, and IaaS examples [2].

Most of the companies are using the combination of all three computing models, and few of the organisations are hiring the developers for PaaS-based applications.

Google Apps, Salesforce, Dropbox, MailChamp, ZenDesk, DocuSign, Slack, Hubspot are the examples of IaaS.

AWS Elastic, Windows Azure, Force.com, OpenShift, Apache Stratos, are the examples of PaaS Cloud Services.

AWS EC2, Google Compute Engine, Digital Ocean, are the examples of SaaS cloud services.

Customers are responsible for protecting their data and monitoring who has access to it in all forms of public cloud services. Cloud storage data protection is critical to effectively implementing and reaping the advantages of cloud computing. Organisations considering common SaaS services such as Microsoft Office 365 or Salesforce should think about how they'll handle their shared responsibility for cloud data security. IaaS providers such as Amazon Web Services (AWS) and Microsoft Azure need a more systematic strategy that begins with data [8].

2.2 Cloud security architecture- Consumer's perspective

Cloud services come in a variety of flavours, including SaaS, PaaS, and IaaS (SPI), using the public, private, and hybrid operating models. The issue and solutions pertaining to Cloud security depends on the patterns. The defined architecture should be aligned with all the issues, and security controls built into the cloud architecture.

So, when designing applications for computing models, what architectural requirement and resources needed for cloud application development and their disposal. In this post, I'll go over how to build "adequate" protection into your IaaS and PaaS applications [9].

2.3 Cloud security model

Let us start with the operational model for cloud protection. In public cloud protection obligations are shared between the cloud service providers and cloud users, while the customer manages all the activities in a private cloud. The shared infrastructure, like routers, is the responsibility of cloud service providers.

Within a cloud service, the figure below depicts the layers that are protected by the provider versus the client (**Figure 5**).

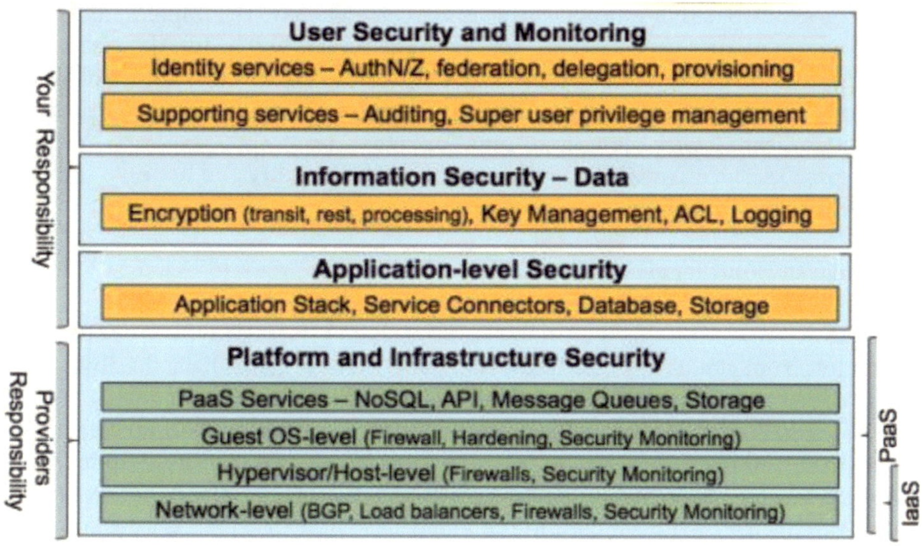

Figure 5.
Layered architecture of Cloud services [5].

It's important to do a difference review on cloud service capabilities before signing up with a provider. This exercise should assess the cloud platform's maturity, accountability, and compliance with enterprise security requirements (such as ISO 27001) as well as regulatory standards like PCI DSS, HIPAA, and SOX. Application migration can be sped up with the use of cloud security maturity models.

The following the security measures are used to ensure the proper safety of the cloud services.

- **Security policies, compliance and practices**: Industry standard frameworks such as ISO 27001, SS 16, and the CSA Cloud controls matrix should be demonstrated by the cloud service provider. Controls approved by the vendor should meet the enterprise data protection standard's control requirements. The scope of controls should be reported when cloud services are approved for ISO 27001 or SSAE 16 [10].

- **Cloud Security architecture:** As per the enterprise norm, the cloud service provider should report security architectural information that either support or impede security management. For example, the virtualization architecture that ensures tenant isolation should be made public.

- **Automation** – Providers can adopt the security automation by publishing API's that used to allow the users to access the logs, privileges and other security threats.

- **Governance and Security Management:** The customer's governance and security management obligations should be specifically defined in comparison to those of the cloud provider.

2.4 Cloud mitigation and security threats

Is cloud computing making the application more vulnerable to security threats? What are the most pressing emerging threats? What are the traditional risks that have been exacerbated or muted? Answers are contingent on the implementation and operating models used by cloud services. The threats will be like data leakage, misconfiguration of services, weakness of VM, attacks via API and failure of VM. The problems can be resolved by making Hardening of VM, incorporating encryption, authentication with security automation, etc. [11].

Threats to service availability, information confidentiality and honesty, must be factored into the design.

2.5 Threat to cloud service availability

DDoS attacks or misconfiguration errors by cloud service providers or customers can interrupt cloud services (SaaS, PaaS, IaaS). These errors have the ability to spread across the cloud, disrupting the network, processes, and storage that are used to host cloud applications. Cloud systems should be designed to survive disturbances to shared resources in order to achieve continuous availability. Applications that were designed to withstand faults within an area, on the other hand, were largely unaffected by the outage and remained accessible to users. As a design philosophy, assume that something will go wrong in the cloud and plan accordingly. Physical hardware failure as well as service interruption within a geographic area should not be a problem for applications.

2.6 Cloud architecture- security services

As a first step, architects should learn about the security features that cloud platforms have (PaaS, IaaS). The framework for integrating protection into cloud services is depicted in the diagram (**Figure 6**).

Offerings and capabilities in terms of security continue to change and differ between cloud providers. As a result, you'll sometimes find that security features like key management and data encryption aren't available. For example, encrypting security objects and keys escrowed to a key management service requires an AES 128 bit encryption service. For such applications that rely on internal resources, a "hybrid cloud" deployment architecture pattern may be the only viable choice. Single Sign-On is another common use case (SSO). If it is a federation architecture

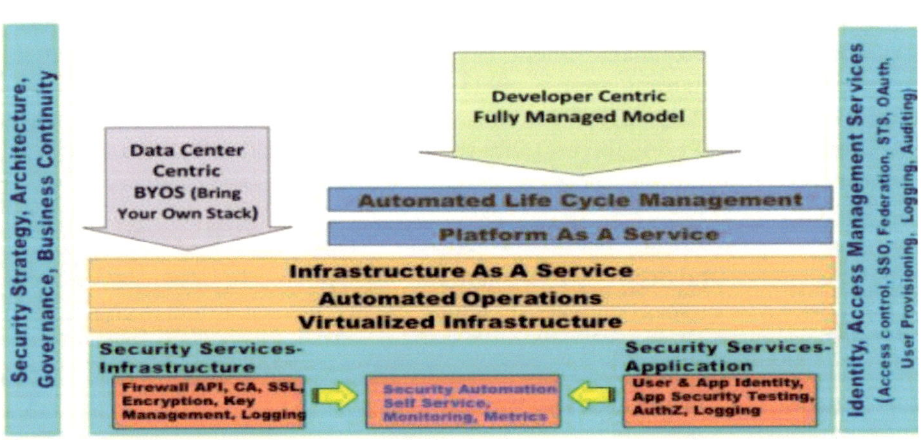

Figure 6.
Cloud security architecture [5].

using SAML 1.1 or 2.0 provided by the cloud service provider, SSO deployed within an organisation may not be extensible to cloud applications.

The following are best practices used in cloud security to mitigate risks to cloud services:

2.6.1 Security service architecture

In the cloud, application implementations necessitate the orchestration of various resources, such as DNS, load balancing, network QoS, and so on. Security automation encompasses the automation of firewall policies between cloud security zones, certificate provisioning (for SSL), virtual machine device configuration, privileged accounts, and log configuration, among other things. Security-related application deployment processes, such as firewall policy development, certificate provisioning, key delivery, and application pen testing, should be moved to a self-service model. This method eliminates human interaction and allows for a security-as-a-service scenario.

2.6.2 Implement identity, access management architecture and practice

Strong user access management infrastructure will be needed as a result of scalable cloud bursting and elastic architecture, which will rely less on network-based access controls. User provisioning and deprovisioning, authentication, federation, authorization, and auditing are all aspects of user and access management lifecycles for both end users and privileged users that should be addressed by cloud access control architecture. In public, private, and hybrid cloud models, a sound architecture would allow reusability of identity and access services for all use cases. Stable token facilities, as well as correct consumer and entitlement provisioning with audit trails, are best practises. The first step in expanding enterprise SSO to cloud services is to construct a federation architecture. Cloud protection partnership is a good place to start.

2.6.3 Automate safeguards

To allow automation, any new security services should be deployed with an API (REST/SOAP). At the time of application deployment, APIs can help simplify firewall rules, configuration hardening, and access control. This can be accomplished by combining open source resources like puppet with the API provided by the cloud service provider.

2.6.4 Encrypt sensitive data

Private cloud applications can be deployed in the public cloud tomorrow. As a result, regardless of the potential operating model, design applications to encrypt all confidential data.

2.6.5 Authenticate IP address and services

Since IP addresses in the cloud are ephemeral, you cannot rely on them to enforce network access control. To allow SSL between cloud providers, use certificates.

2.6.6 Log, log, log

All security activities should be logged centrally in order to build an end-to-end transaction view of non-repudiation characteristics. Logs and audit trails are the only accurate evidence used by forensic engineers to analyse and understand how an application was compromised in the case of a security incident.

2.6.7 Continuously monitor cloud services

Given that preventive controls cannot meet all enterprise requirements, monitoring is an essential feature. To perform security event correlation, security monitoring should use logs produced by cloud services, APIs, and hosted cloud applications. The CSA's cloud audit (cloudaudit.org) will help with this mission.

2.7 Cloud security principles

The product development culture, emerging technology implementation, IT service delivery models, technology policy, and investments made in the field of security tools and capabilities all show that each company has a different level of risk tolerance. When a company's business unit chooses to use SaaS for business purposes, the technology architecture changes. The security architecture should also be consistent with the technology architecture and principles. An enterprise technology architect should understand and configure the following cloud security concepts [12]:

- Cloud-based services should adhere to the concept of least privilege.
- Using firewalls and container – isolation between different protection zones should be ensured. Cloud firewall policies should adhere to data sensitivity-based trust zone isolation requirements.
- End-to-end transport level encryption (SSL, TLS, IPSEC) can be used by applications to protect data in transit between cloud and business applications.
- Authentication and authorization should be delegated to trusted security providers by applications. SAML 2.0 can be used to support single sign-on.
- Enterprise standard VM images can be used to deploy applications in a trusted zone.
- When implementing a virtual private cloud, industry standard VPN protocols including SSL, SSH, etc.
- Using an API, cloud security monitoring can be combined with existing security tools and services.

2.8 Cloud security architecture patterns

Cloud security risks can be mitigated by designing effective security controls that secure the CIA of information in the cloud. The vendor, the enterprise, or a third-party provider can provide security controls as a service (Security-as-a-Service). The point of security controls (safeguards) – technologies and processes – is usually where security architectural trends are expressed [13].

Security architecture trends act as a compass, allowing developers to move applications to the cloud faster while minimising security risks. Furthermore, cloud security architecture trends should emphasise the trust boundary between different cloud services and components. Normal interfaces and protection protocols (SSL, TLS, IPSEC, LDAPS, and so on) should also be highlighted in these patterns.

Finally, the patterns can be used to build security checklists that can be automated using configuration management software such as puppet.

For each of the security resources consumed by the cloud application, trends should highlight the following attributes (but not limited to):

- Logical place – in-house, third-party cloud, native to cloud service The service's efficiency, availability, firewall policy, and governance can all be affected by its venue.

- Protocol(s) – Which protocol(s) is/are used to call the service? For e.g., for service requests, REST with X.509 certificates.

- Service feature – What is the service's purpose?

- Input/output – What are the security service's inputs, including monitoring methods and outputs? Input = XML doc and Output = XML doc with encrypted attributes, for example.

- Overview of the security control – What security controls does the security service provide? For instance, information confidentiality at rest, user authentication, and device authentication.

2.8.1 Security services based on infrastructure

A cloud service provider is required to provide security controls for DoS privacy, as well as confidentiality and integrity protection for sessions originating from mobile and PC, according to the pattern.

2.8.2 Application based security services

Identification, authentication, access enforcement, system identification, cryptographic services, and key management can all be handled by the cloud service provider, the corporate data centre, or a combination of the two (**Figure 7**).

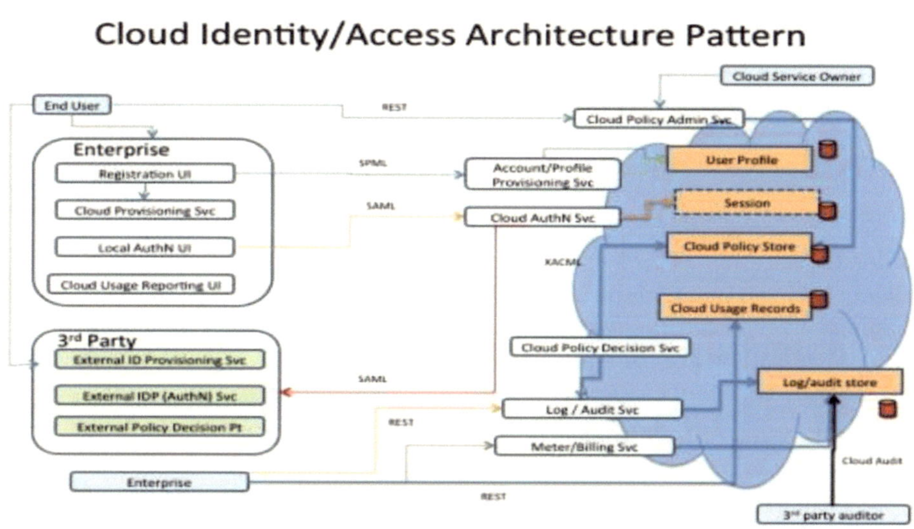

Figure 7.
Identity and access pattern [5].

User registration, authentication, account provisioning, policy compliance, logging, auditing, and metering are all examples of typical cloud access control use cases illustrated in this pattern. It focuses on the actors who communicate with cloud, in-house (enterprise), and third-party hosted services:

2.9 Identifications of security services

- An authentication service that allows users to log in via an enterprise portal (Local AuthN UI) and is usually provided via the SAML protocol. A cloud session store stores the authenticated session state.

- The account and profile provisioning service facilitates the development of new accounts and user profiles, usually through the use of SPML (Service Provisioning Markup Language).

- The cloud policy admin service is used to manage policies that control which cloud services end users have access to. Cloud service owners (enterprises) can use this service to perform administrative tasks, while end users can request access to cloud services. The cloud policy store is where cloud policies are held.

- The logging and auditing service can be used for two purposes. The first is cloud-based event reporting, which includes security events, and the second is auditing. This service can be accessed using Cloud Audit protocols and APIs.

- The metering programme keeps track of how much cloud resources are being used. This service can be used for chargebacks as well as billing reconciliation by finance departments.

2.10 Identity security services in the Enterprise

- Domain registration UI is a user interface for registering, managing, and provisioning new cloud services. The cloud providers implement authentication and authorization.

- End users produce usage reports using the cloud usage reporting UI.

- A cloud provisioning service is used to make cloud resources accessible (compute, storage, network, application services).

2.11 Third party identification of security services

Identity services provided by a third party and hosted at their location are used by cloud applications. Third-party users who need access to cloud infrastructure to conduct business functions on behalf of the company may get help from services. Backup and device control, for example. The third-party provider is in charge of user authentication, provisioning and access enforcement.

3. Cloud security challenges

According to Gartner, the global public cloud services market will increase 17 percent to $266.4 billion in 2020, up from $227.8 billion in 2019. In its study "high risk to Cloud Computing: Egregious Eleven," the CSA (cloud security alliance) outlined the following major cloud challenges.

3.1 Data breaches

The effects of data breaches will include the following:

- Impact on customer or partner credibility and confidence
- Loss of intellectual property (IP) to rivals, which could have an effect on the release of goods.
- Regulatory ramifications that could lead to financial loss.
- Brand effect, which may result in a decrease in market value due to the factors mentioned above.
- Legislative and contractual obligations.
- Expenses incurred as a result of incident response and forensics

3.2 Improper change management

This is one of the most popular cloud challenges. In 2017, a misconfigured AWS Simple Storage Service (S3) cloud storage bucket exposed 123 million American households' detailed and private data. Experian, a credit bureau, owned the data collection, which it sold to Alteryx, an online marketing and data analytics firm. Such occurrences have the potential to be catastrophic.

3.3 Poor architecture and mechanism

Organisations all over the world are moving parts of their IT infrastructure to public clouds. The introduction of sufficient security infrastructure to withstand cyberattacks is one of the most difficult challenges during this transition. Unfortunately, many businesses are still baffled by this operation. Another contributing factor is a lack of awareness of the shared security obligation model.

3.4 Improper identification and key management

Multiple improvements to standard internal system management procedures related to identity and access management are introduced by cloud computing (IAM). These aren't even brand-new problems. Rather, when dealing with the cloud, they are more serious concerns because cloud computing has a significant effect on identity, certificate, and access management.

3.5 Threats within the organisation

The employee within the organisation can be a threat by using the sensitive data in their personal use or sharing the confidential data with others.

3.6 Wrong interfaces and API

Customers can manage and communicate with cloud services through a series of software user interfaces (UIs) and APIs exposed by cloud computing providers. The security and availability of general cloud services are also reliant on these APIs' security. APIs that aren't well-designed can lead to misuse or, worse, a data breach.

APIs that have been broken, leaked, or compromised have resulted in significant data breaches.

4. Cloud security solutions

To address the primary cloud security challenges in terms of visibility and control the following requirement need to be accessed [14, 15].

4.1 Access to the cloud service

Direct access to the cloud service is needed for a full view of cloud data. An application programming interface (API) access to the cloud service is used by cloud security solutions to achieve this. It is possible to access data using an API link:

- Where does your data go in the cloud?
- Who is making use of cloud data?
- The positions of consumers of cloud service access.
- With whom do cloud users share data?
- The location of cloud data.
- The location from which cloud data is accessed and downloaded, as well as the user.

4.2 Cloud data control

Need to apply the controls that best suit the organisations demands. These safeguards include:

- Classification — As data is generated in the cloud, classify it on multiple levels, such as confidential, controlled, or public. Data may be prohibited from entering or exiting the cloud service after it has been classified.
- Implement a cloud data loss prevention (DLP) solution to protect data from unauthorised access and automatically disable access and data transport when suspicious behaviour is detected.
- Manage collaboration controls in the cloud service, such as reducing file and folder permissions for specific users to editor or viewer, deleting permissions, and revoking shared links. Ensure the cloud data should be encrypted from unauthorised users.

4.2.1 Data access and its applications

Security relies heavily on access control.

- User access control — Set up device and application access controls to ensure that only approved users have access to cloud data and applications. To implement access controls, a Cloud Access Security Broker (CASB) may be used.

- User access control — If a personal, unauthorised device attempts to access cloud data, access is denied.

- Malicious behaviour detection — Use user behaviour analytics (UBA) to detect compromised accounts and insider attacks, preventing malicious data exfiltration.

- Malware protection — Use techniques like file inspection, device whitelisting, machine learning-based malware identification, and network traffic analysis to keep malware out of cloud services.

4.3 Compliance standards and policies

The policies and standards should be updated and expanded as per the current and forthcoming threats.

- Risk assessment — Re-evaluate and upgrade risk assessments to incorporate cloud services. Identify and mitigate the risks posed by cloud environments and providers. To speed up the evaluation process, risk databases for cloud providers are available.

- Application regulatory requirements like PCI, HIPAA, Sarbanes-Oxley, etc. and its assessment.

4.4 Cloud security importance

According to news reports, one out of every four businesses that use public cloud services has had data stolen by a malicious actor. An additional one out of every five people has had an advanced assault on their public cloud infrastructure. According to the same survey, 83 percent of businesses said they store confidential data in the cloud. With 97 percent of businesses using cloud services today, it's critical that everyone assesses their cloud security and establishes a data-protection strategy.

McAfee's cloud protection helps businesses grow faster by allowing them complete visibility and control over their data in the cloud. Find out more about McAfee's cloud protection technologies.

Cloud MVISION

The enterprise's multi-cloud protection platform. Create a single protection policy that can be used through SaaS, PaaS, IaaS, Containers, and the Web. Accelerate cloud adoption by simplifying security for a distributed workforce.

Unified Cloud Edge (UCE) is a component of MVISION Cloud that integrates data security from devices, the network, and the cloud to make SASE architecture adoption easier.

Platform for Cloud-Native Application Security (CNAPP).

CNAPP, which is part of MVISION Cloud, audits and secures the entire IaaS/PaaS stack, including containers and private clouds.

5. Conclusion

We can integrate protection into your software without having to reinvent the wheel inside your app's boundaries, saving money on "bolt-on" safeguards. Creating security standards and architectural patterns that can be used in the design process is a good practise. During the design process, architectural trends will assist in articulating where controls are applied (Cloud versus third party versus enterprise) so that sufficient security controls are baked into the application design. When designing cloud protection trends, keep in mind the applicable threats and the risk-appropriate principle. Discussed the various security challenges and its possible security solutions that mostly needed for the secure system. Finally, a cloud protection architecture should meet the needs of developers in terms of protecting the confidentiality, integrity, and availability of data processed and stored in the cloud.

Author details

Jagdish Chandra Patni
School of Computer Science, University of Petroleum and Energy Studies, Dehradun, India

*Address all correspondence to: patnijack@gmail.com

References

[1] Move confidently to hybrid multicloud and integrate security into every phase of your cloud journey, The premier hybrid cloud and AI event May 11-May 12, Americas

[2] The Different type of cloud computing and how they differ, https://www.vxchnge.com/ (April 2021)

[3] https://www.citrix.com/en-in/glossary/what-is-public-cloud.html (March 2021)

[4] Six Common Challenges of Cloud Implementations, white paper, September 2014.

[5] Managed Cloud Services for Hyper Performance and Uninterrupted Continuity, Cloud 4C, CtrlS (March 2021)

[6] Introduction to Cloud Security Architecture from a Cloud Consumer's Perspective by Subra Kumaraswamy, InfoQ, Dec 07, 2011

[7] The cloud-based energy and asset management platform from Siemens powered by MindSphere

[8] Roshana Gul, The Relationship between Reputation, Customer Satisfaction, Trust, and Loyalty, Journal of Public Administration and Governance ISSN 2161-7104 2014, Vol. 4, No. 3, September 22, 2014

[9] Jaydip Sen, Security and Security and Privacy Privacy Privacy Issues in Cloud Computing Computing, Innovation Labs, Tata Consultancy Services Ltd., Kolkata, INDIA

[10] Security for Cloud Computing Ten Steps to Ensure Success, Cloud Standards Customer Council, 2017

[11] What is cloud Security, https://www.mcafee.com/ (May 2021).

[12] Cloud Security Challenges, https://cloudsecurityalliance.fr/ (April 2021)

[13] Hybrid Cloud, https://www.netapp.com/hybrid-cloud/what-is-hybrid-cloud/ (April 2021)

[14] Wissam Razouk, Daniele Sgandurra, and Kouichi Sakurai. 2017. A new security middleware architecture based on fog computing and cloud to support IoT constrained devices. In Proceedings of the 1st International Conference on Internet of Things and Machine Learning (IML '17). Association for Computing Machinery, New York, NY, USA, Article 35, 1-8.

[15] J. Yang, L. Zhang and X. A. Wang, "On Cloud Computing Middleware Architecture," *2015 10th International Conference on P2P, Parallel, Grid, Cloud and Internet Computing (3PGCIC)*, 2015, pp. 832-835, doi: 10.1109/3PGCIC.2015.46.